September 2012

OIL AND GAS

Information on Shale Resources, Development, and Environmental and Public Health Risks

OIL AND GAS

Information on Shale Resources, Development, and Environmental and Public Health Risks

Highlights of GAO-12-732, a report to congressional requesters

Why GAO Did This Study

New applications of horizontal drilling techniques and hydraulic fracturing—in which water, sand, and chemical additives are injected under high pressure to create and maintain fractures in underground formations—allow oil and natural gas from shale formations (known as "shale oil" and "shale gas") to be developed. As exploration and development of shale oil and gas have increased—including in areas of the country without a history of oil and natural gas development—questions have been raised about the estimates of the size of these resources, as well as the processes used to extract them.

GAO was asked to determine what is known about the (1) size of shale oil and gas resources and the amount produced from 2007 through 2011 and (2) environmental and public health risks associated with the development of shale oil and gas. GAO reviewed estimates and data from federal and nongovernmental organizations on the size and production of shale oil and gas resources. GAO also interviewed federal and state regulatory officials, representatives from industry and environmental organizations, oil and gas operators, and researchers from academic institutions.

GAO is not making any recommendations in this report. We provided a draft of this report to the Department of Energy, the Department of the Interior, and the Environmental Protection Agency for review. The Department of the Interior and the Environmental Protection Agency provided technical comments, which we incorporated as appropriate. The Department of Energy did not provide comments.

View GAO-12-732. For more information, contact Frank Rusco at (202) 512-3841 or ruscof@gao.gov.

What GAO Found

Estimates of the size of shale oil and gas resources in the United States by the Energy Information Administration (EIA), U.S. Geological Survey (USGS), and the Potential Gas Committee—three organizations that estimate the size of these resources—have increased over the last 5 years, which could mean an increase in the nation's energy portfolio. For example, in 2012, EIA estimated that the amount of technically recoverable shale gas in the United States was 482 trillion cubic feet—an increase of 280 percent from EIA's 2008 estimate. However, according to EIA and USGS officials, estimates of the size of shale oil and gas resources in the United States are highly dependent on the data, methodologies, model structures, and assumptions used to develop them. In addition, less is known about the amount of technically recoverable shale oil than shale gas, in part because large-scale production of shale oil has been under way for only the past few years. Estimates are based on data available at a given point in time and will change as additional information becomes available. In addition, domestic shale oil and gas production has experienced substantial growth; shale oil production increased more than fivefold from 2007 to 2011, and shale gas production increased more than fourfold from 2007 to 2011.

Oil and gas development, whether conventional or shale oil and gas, pose inherent environmental and public health risks, but the extent of these risks associated with shale oil and gas development is unknown, in part, because the studies GAO reviewed do not generally take into account the potential long-term, cumulative effects. For example, according to a number of studies and publications GAO reviewed, shale oil and gas development poses risks to air quality, generally as the result of (1) engine exhaust from increased truck traffic, (2) emissions from diesel-powered pumps used to power equipment, (3) gas that is flared (burned) or vented (released directly into the atmosphere) for operational reasons, and (4) unintentional emissions of pollutants from faulty equipment or impoundments—temporary storage areas. Similarly, a number of studies and publications GAO reviewed indicate that shale oil and gas development poses risks to water quality from contamination of surface water and groundwater as a result of erosion from ground disturbances, spills and releases of chemicals and other fluids, or underground migration of gases and chemicals. For example, tanks storing toxic chemicals or hoses and pipes used to convey wastes to the tanks could leak, or impoundments containing wastes could overflow as a result of extensive rainfall. According to the New York Department of Environmental Conservation's 2011 Supplemental Generic Environmental Impact Statement, spilled, leaked, or released chemicals or wastes could flow to a surface water body or infiltrate the ground, reaching and contaminating subsurface soils and aquifers. In addition, shale oil and gas development poses a risk to land resources and wildlife habitat as a result of constructing, operating, and maintaining the infrastructure necessary to develop oil and gas; using toxic chemicals; and injecting fluids underground. However, the extent of these risks is unknown. For example, the studies and publications GAO reviewed on air quality conditions provide information for a specific site at a specific time but do not provide the information needed to determine the overall cumulative effects that shale oil and gas activities may have on air quality. Further, the extent and severity of environmental and public health risks identified in the studies and publications GAO reviewed may vary significantly across shale basins and also within basins because of location- and process-specific factors, including the location and rate of development; geological characteristics, such as permeability, thickness, and porosity of the formations; climatic conditions; business practices; and regulatory and enforcement activities.

_____ **United States Government Accountability Office**

Contents

Abbreviations

BLM	Bureau of Land Management
Btu	British thermal unit
DOE	Department of Energy
EIA	Energy Information Administration
EPA	Environmental Protection Agency
NORM	naturally occurring radioactive materials
Tcf	technically recoverable gas
USGS	U.S. Geological Survey

United States Government Accountability Office
Washington, DC 20548

September 5, 2012

Congressional Requesters

For decades, the United States has relied on imports of oil and natural gas to meet domestic needs. As recently as 2007, the expectation was that the nation would increasingly rely on imports of natural gas to meet its growing demand. However, recent improvements in technology have allowed companies that develop petroleum resources to extract oil and natural gas from shale formations,[1] known as "shale oil" and "shale gas," respectively, which were previously inaccessible because traditional techniques did not yield sufficient amounts for economically viable production. In particular, as we reported in January 2012, new applications of horizontal drilling techniques and hydraulic fracturing—a process that injects a combination of water, sand, and chemical additives under high pressure to create and maintain fractures in underground rock formations that allow oil and natural gas to flow—have prompted a boom in shale oil and gas production.[2] According to the Department of Energy (DOE), America's shale gas resource base is abundant, and development of this resource could have beneficial effects for the nation, such as job creation.[3] According to a report by the Baker Institute, domestic shale gas development could limit the need for expensive imports of these resources—helping to reduce the U.S. trade deficit.[4] In addition, replacing older coal burning power generation with new natural gas-fired generators could reduce greenhouse gas emissions and result in fewer air pollutants

---

[1]Shale oil differs from "oil shale." Shale is a sedimentary rock that is predominantly composed of consolidated clay-sized particles. Oil shale requires a different process to extract. Specifically, to extract the oil from oil shale, the rock needs to be heated to very high temperatures—ranging from about 650 to 1,000 degrees Fahrenheit—in a process known as retorting. Oil shale is not currently economically viable to produce. For additional information on oil shale, see GAO, *Energy-Water Nexus: A Better and Coordinated Understanding of Water Resources Could Help Mitigate the Impacts of Potential Oil Shale Development*, GAO-11-35 (Washington, D.C.: Oct. 29, 2010).

[2]GAO, *Energy-Water Nexus: Information on the Quantity, Quality, and Management of Water Produced during Oil and Gas Production*, GAO-12-156 (Washington, D.C.: Jan. 9, 2012).

[3]EIA is a statistical agency within DOE that provides independent data, forecasts, and analyses.

[4]The Baker Institute is a public policy think tank located on the Rice University campus.

for the same amount of electric power generated.[5] Early drilling activity in shale formations was centered primarily on natural gas, but with the falling price of natural gas companies switched their focus to oil and natural gas liquids, which are a more valuable product.[6]

As exploration and development of shale oil and gas have increased in recent years—including in areas of the country without a history of oil and natural gas activities—questions have been raised about the estimates of the size of domestic shale oil and gas resources, as well as the processes used to extract them.[7] For example, some organizations have questioned the accuracy of the estimates of the shale gas supply. In particular, some news organizations have reported concerns that such estimates may be inflated. In addition, concerns about environmental and public health effects of the increased use of horizontal drilling and hydraulic fracturing, particularly on air quality and water resources, have garnered extensive public attention. According to the International Energy Agency, some questions also exist about whether switching from coal to natural gas will lead to a reduction in greenhouse gas emissions—based, in part, on uncertainty about additional emissions from the development of shale gas. These concerns and other considerations have led some communities and certain states to impose restrictions or moratoriums on drilling operations to allow time to study and better understand the potential risks associated with these practices.

In this context, you asked us to provide information on shale oil and gas. This report describes what is known about (1) the size of shale oil and gas resources in the United States and the amount produced from 2007 through 2011—the years for which data were available—and (2) the environmental and public health risks associated with development of shale oil and gas.[8]

---

[5]EIA reported that using natural gas over coal would lower emissions in the United States, but some researchers have reported that greater reliance on natural gas would fail to significantly slow climate change.

[6]The natural gas liquids include propane, butane, and ethane, and are separated from the produced gas at the surface in lease separators, field facilities, or gas processing plants.

[7]For the purposes of this report, resources represent all oil or natural gas contained within a formation and can be divided into resources and reserves.

[8]For the purposes of this report, we refer to risk as a threat or vulnerability that has potential to cause harm.

To determine what is known about the size of shale oil and gas resources and the amount of shale oil and gas produced, we collected data from federal agencies, state agencies, private industry, and academic organizations. Specifically, to determine what is known about the size of these resources, we obtained information for technically recoverable and proved reserves estimates for shale oil and gas from the EIA, the U.S. Geological Survey (USGS), and the Potential Gas Committee—a nongovernmental organization composed of academics and industry representatives. We interviewed key officials from these agencies and the committee about the assumptions and methodologies used to estimate the resource size. Estimates of proved reserves of shale oil and gas are based on data provided to EIA by operators—companies that develop petroleum resources to extract oil and natural gas.[9] To determine what is known about the amount of shale oil and gas produced from 2007 through 2011, we obtained data from EIA—which is responsible for estimating and reporting this and other energy information. To assess the reliability of these data, we examined EIA's published methodology for collecting this information and interviewed key EIA officials regarding the agency's data collection efforts. We also met with officials from states, representatives from private industry, and researchers from academic institutions who are familiar with these data and EIA's methodology. We discussed the sources and reliability of the data with these officials and found the data sufficiently reliable for the purposes of this report. For all estimates we report, we reviewed the methodologies used to derive them and also found them sufficiently reliable for the purposes of this report.

To determine what is known about the environmental and public health risks associated with the development of shale oil and gas,[10] we reviewed studies and other publications from federal agencies and laboratories, state agencies, local governments, the petroleum industry, academic institutions, environmental and public health groups, and other nongovernmental associations. We identified these studies by conducting

---

[9]Proved reserves refer to the amount of oil and gas that have been discovered and defined.

[10]Operators may use hydraulic fracturing to develop oil and natural gas from formations other than shale, but for the purposes of this report we focused on development of shale formations. Specifically, coalbed methane and tight sandstone formations may rely on these practices and some studies and publications we reviewed identified risks that can apply to these formations. However, many of the studies and publications we identified and reviewed focused primarily on shale formations.

a literature search, and by asking for recommendations during interviews with federal, state, and tribal officials; representatives from industry, trade organizations, environmental, and other nongovernmental groups; and researchers from academic institutions. For a number of studies, we interviewed the author or authors to discuss the study's findings and limitations, if any. We believe we have identified the key studies through our literature review and interviews, and that the studies included in our review have accurately identified currently known potential risks for shale oil and gas development. However, it is possible that we may not have identified all of the studies with findings relevant to our objectives, and the risks we present may not be the only issues of concern.

The risks identified in the studies and publications we reviewed cannot, at present, be quantified, and the magnitude of potential adverse affects or likelihood of occurrence cannot be determined for several reasons. First, it is difficult to predict how many or where shale oil and gas wells may be constructed. Second, the extent to which operators use effective best management practices to mitigate risk may vary. Third, based on the studies we reviewed, there are relatively few studies that are based on comparing predevelopment conditions to postdevelopment conditions— making it difficult to detect or attribute adverse conditions to shale oil and gas development. In addition, changes to the federal, state, and local regulatory environments and the effectiveness of implementing and enforcing regulations will affect operators' future activities and, therefore, the level of risk associated with future development of oil and gas resources. Moreover, risks of adverse events, such as spills or accidents, may vary according to business practices which, in turn, may vary across oil and gas companies, making it difficult to distinguish between risks associated with the process to develop shale oil and gas from risks that are specific to particular business practices. To obtain additional perspectives on issues related to environmental and public health risks, we interviewed federal officials from DOE's National Energy Technical Laboratory, the Department of the Interior's Bureau of Land Management (BLM) and Bureau of Indian Affairs, and the Environmental Protection Agency (EPA); state regulatory officials from Arkansas, Colorado, Louisiana, North Dakota, Ohio, Oklahoma, Pennsylvania, and Texas;[11] tribal officials from the Osage Nation; shale oil and gas operators;

---

[11]We selected these states because they are involved with shale oil and gas development.

representatives from environmental and public health organizations; and other knowledgeable parties with experience related to shale oil and gas development, such as researchers from the Colorado School of Mines, the University of Texas, Oklahoma University, and Stanford University. Appendix I provides additional information on our scope and methodology.

We conducted this performance audit from November 2011 to September 2012 in accordance with generally accepted government auditing standards. Those standards require that we plan and perform the audit to obtain sufficient, appropriate evidence to provide a reasonable basis for our findings and conclusions based on our audit objectives. We believe that the evidence obtained provides a reasonable basis for our findings and conclusions based on our audit objectives.

Background

This section includes (1) an overview of oil and natural gas, (2) the shale oil and gas development process, (3) the regulatory framework, (4) the location of shale oil and gas in the United States, and (5) information on estimating the size of these resources.

Overview

Oil and natural gas are found in a variety of geologic formations. Conventional oil and natural gas are found in deep, porous rock or reservoirs and can flow under natural pressure to the surface after drilling. In contrast to the free-flowing resources found in conventional formations, the low permeability of some formations, including shale, means that oil and gas trapped in the formation cannot move easily within the rock. On one extreme—oil shale, for example—the hydrocarbon trapped in the shale will not reach a liquid form without first being heated to very high temperatures—ranging from about 650 to 1,000 degrees Fahrenheit—in a process known as retorting. In contrast, to extract shale oil and gas from the rock, fluids and proppants (usually sand or ceramic beads used to hold fractures open in the formation) are injected under high pressure to create and maintain fractures to increase permeability, thus allowing oil or gas to be extracted. Other formations, such as coalbed methane formations and tight sandstone formations,[12] may also require stimulation to allow oil or gas to be extracted.[13]

Most of the energy used in the United States comes from fossil fuels such as oil and natural gas. Oil supplies more than 35 percent of all the energy the country consumes, and almost the entire U.S. transportation fleet—cars, trucks, trains, and airplanes—depends on fuels made from oil. Natural gas is an important energy source to heat buildings, power the industrial sector, and generate electricity. Natural gas provides more than 20 percent of the energy used in the United States,[14] supplying nearly half of all the energy used for cooking, heating, and powering other home appliances, and generating almost one-quarter of U.S. electricity supplies.

The Shale Oil and Gas Development Process

The process to develop shale oil and gas is similar to the process for conventional onshore oil and gas, but shale formations may rely on the use of horizontal drilling and hydraulic fracturing—which may or may not be used on conventional wells. Horizontal drilling and hydraulic fracturing are not new technologies, as seen in figure 1, but advancements, refinements, and new uses of these technologies have greatly expanded oil and gas operators' abilities to use these processes to economically develop shale oil and gas resources. For example, the use of multistage hydraulic fracturing within a horizontal well has only been widely used in the last decade.[15]

---

[12]Conventional sandstone has well-connected pores, but tight sandstone has irregularly distributed and poorly connected pores. Due to this low connectivity or permeability, gas trapped within tight sandstone is not easily produced.

[13]For coalbed methane formations, the reduction in pressure needed to extract gas is achieved through dewatering. As water is pumped out of the coal seams, reservoir pressure decreases, allowing the natural gas to release (desorb) from the surface of the coal and flow through natural fracture networks into the well.

[14]Ground Water Protection Council and ALL Consulting, *Modern Shale Gas Development in the United States: A Primer*, a special report prepared at the request of the Department of Energy (Washington, D.C.: April 2009).

[15]Hydraulic fracturing is often conducted in stages. Each stage focuses on a limited linear section and may be repeated numerous times.

Figure 1: History of Horizontal Drilling and Hydraulic Fracturing

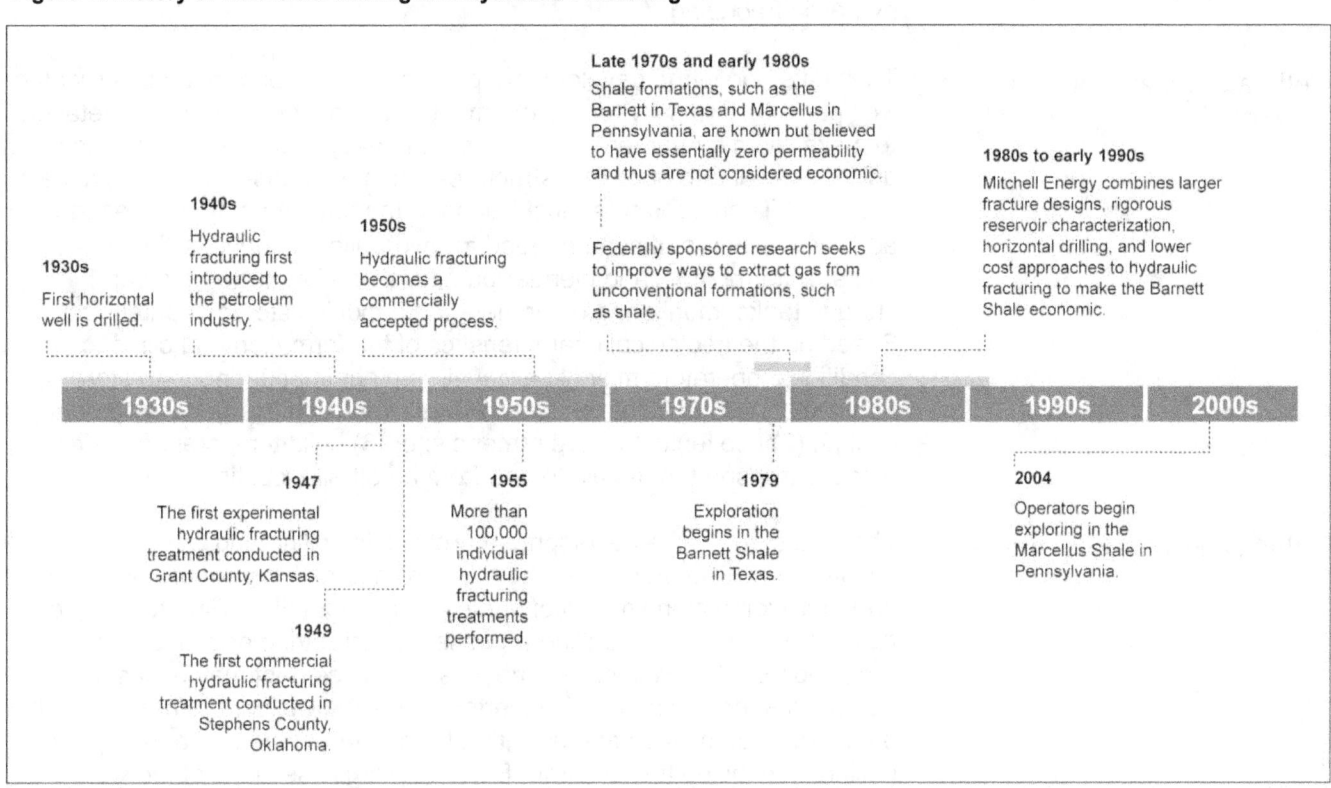

Late 1970s and early 1980s
Shale formations, such as the Barnett in Texas and Marcellus in Pennsylvania, are known but believed to have essentially zero permeability and thus are not considered economic.

Federally sponsored research seeks to improve ways to extract gas from unconventional formations, such as shale.

1980s to early 1990s
Mitchell Energy combines larger fracture designs, rigorous reservoir characterization, horizontal drilling, and lower cost approaches to hydraulic fracturing to make the Barnett Shale economic.

1940s
Hydraulic fracturing first introduced to the petroleum industry.

1950s
Hydraulic fracturing becomes a commercially accepted process.

1930s
First horizontal well is drilled.

1930s	1940s	1950s	1970s	1980s	1990s	2000s

1947
The first experimental hydraulic fracturing treatment conducted in Grant County, Kansas.

1949
The first commercial hydraulic fracturing treatment conducted in Stephens County, Oklahoma.

1955
More than 100,000 individual hydraulic fracturing treatments performed.

1979
Exploration begins in the Barnett Shale in Texas.

2004
Operators begin exploring in the Marcellus Shale in Pennsylvania.

Source: GAO.

First, operators locate suitable shale oil and gas targets using seismic methods of exploration,[16] negotiate contracts or leases that allow mineral development, identify a specific location for drilling, and obtain necessary permits; then, they undertake a number of activities to develop shale oil and gas. The specific activities and steps taken to extract shale oil and gas vary based on the characteristics of the formation, but the development phase generally involves the following stages: (1) well pad

---

[16]The seismic method of exploration introduces energy into the subsurface through explosions in shallow "shot holes" by striking the ground forcefully (with a truck-mounted thumper), or by vibration methods. A portion of the energy returns to the surface after being reflected from the subsurface strata. This energy is detected by surface instruments, called geophones, and the information carried by the energy is processed by computers to interpret subsurface conditions.

GAO-12-732 Shale Oil and Gas Development

preparation and construction, (2) drilling and well construction, and (3) hydraulic fracturing.[17]

Well Pad Preparation and Construction

The first stage in the development process is to prepare and construct the well pad site. Typically, operators must clear and level surface vegetation to make room for numerous vehicles and heavy equipment—such as the drilling rig—and to build infrastructure—such as roads—needed to access the site.[18] Then operators must transport the equipment that mixes the additives, water, and sand needed for hydraulic fracturing to the site—tanks, water pumps, and blender pumps, as well as water and sand storage tanks, monitoring equipment, and additive storage containers . Based on the geological characteristics of the formation and climatic conditions, operators may (1) excavate a pit or impoundment to store freshwater, drilling fluids, or drill cuttings—rock cuttings generated during drilling; (2) use tanks to store materials; or (3) build temporary transfer pipes to transport materials to and from an off-site location.

Drilling and Well Construction

The next stage in the development process is drilling and well construction. Operators drill a hole (referred to as the wellbore) into the earth through a combination of vertical and horizontal drilling techniques. At several points in the drilling process, the drill string and bit are removed from the wellbore so that casing and cement may be inserted. Casing is a metal pipe that is inserted inside the wellbore to prevent high-pressure fluids outside the formation from entering the well and to prevent drilling mud inside the well from fracturing fragile sections of the wellbore. As drilling progresses with depth, casings that are of a smaller diameter than the hole created by the drill bit are inserted into the wellbore and bonded in place with cement, sealing the wellbore from the surrounding formation.

Drilling mud (a lubricant also known as drilling fluid) is pumped through the wellbore at different densities to balance the pressure inside the wellbore and bring rock particles and other matter cut from the formation back to the rig. A blowout preventer is installed over the well as a safety measure to prevent any uncontrolled release of oil or gas and help

---

[17]The specific order of activities and steps may vary.

[18]According to the New York Department of Environmental Conservation's 2011 Supplemental Generic Environmental Impact Statement, the average size of a well pad is 3.5 acres.

maintain control over pressures in the well. Drill cuttings, which are made up of ground rock coated with a layer of drilling mud or fluid, are brought to the surface. Mud pits provide a reservoir for mixing and holding the drilling mud. At the completion of drilling, the drilling mud may be recycled for use at another drilling operation.

Instruments guide drilling operators to the "kickoff point"—the point that drilling starts to turn at a slight angle and continues turning until it nears the shale formation and extends horizontally. Production casing and cement are then inserted to extend the length of the borehole to maintain wellbore integrity and prevent any communication between the formation fluids and the wellbore. After the casing is set and cemented, the drilling operator may run a cement evaluation log by lowering an electric probe into the well to measure the quality and placement of the cement. The purpose of the cement evaluation log is to confirm that the cement has the proper strength to function as designed—preventing well fluids from migrating outside the casing and infiltrating overlying formations. After vertical drilling is complete, horizontal drilling is conducted by slowly angling the drill bit until it is drilling horizontally. Horizontal stretches of the well typically range from 2,000 to 6,000 feet long but can be as long as 12,000 feet long, in some cases.

Throughout the drilling process, operators may vent or flare some natural gas, often intermittently, in response to maintenance needs or equipment failures. This natural gas is either released directly into the atmosphere (vented) or burned (flared). In October 2010, we reported on venting and flaring of natural gas on public lands.[19] We reported that vented and flared gas on public lands represents potential lost royalties for the federal government and contributes to greenhouse gas emissions. Specifically, venting releases methane and volatile organic compounds, and flaring emits carbon dioxide, both greenhouse gases that contribute to global climate change. Methane is a particular concern since it is a more potent greenhouse gas than carbon dioxide.

Hydraulic Fracturing

The next stage in the development process is stimulation of the shale formation using hydraulic fracturing. Before operators or service companies perform a hydraulic fracture treatment of a well, a series of

---

[19]GAO, *Federal Oil and Gas Leases: Opportunities Exist to Capture Vented and Flared Natural Gas, Which Would Increase Royalty Payments and Reduce Greenhouse Gases*, GAO-11-34 (Washington, D.C.: Oct. 29, 2010).

tests may be conducted to ensure that the well, wellhead equipment, and fracturing equipment can safely withstand the high pressures associated with the fracturing process. Minimum requirements for equipment pressure testing can be determined by state regulatory agencies for operations on state or private lands. In addition, fracturing is conducted below the surface of the earth, sometimes several thousand feet below, and can only be indirectly observed. Therefore, operators may collect subsurface data—such as information on rock stresses[20] and natural fault structures—needed to develop models that predict fracture height, length, and orientation prior to drilling a well. The purpose of modeling is to design a fracturing treatment that optimizes the location and size of induced fractures and maximizes oil or gas production.

To prepare a well to be hydraulically fractured, a perforating tool may be inserted into the casing and used to create holes in the casing and cement. Through these holes, fracturing fluid—that is injected under high pressures—can flow into the shale (fig. 2 shows a used perforating tool).

---

[20]Stresses in the formation generally define a maximum and minimum stress direction that influence the direction a fracture will grow.

Figure 2: Perforating Tool

Source: GAO.

Fracturing fluids are tailored to site specific conditions, such as shale thickness, stress, compressibility, and rigidity. As such, the chemical additives used in a fracture treatment vary. Operators may use computer models that consider local conditions to design site-specific hydraulic fluids. The water, chemicals, and proppant used in fracturing fluid are typically stored on-site in separate tanks and blended just before they are injected into the well. Figure 3 provides greater detail about some chemicals commonly used in fracturing.

GAO-12-732 Shale Oil and Gas Development

Figure 3: Examples of Common Ingredients Found in Fracturing Fluid

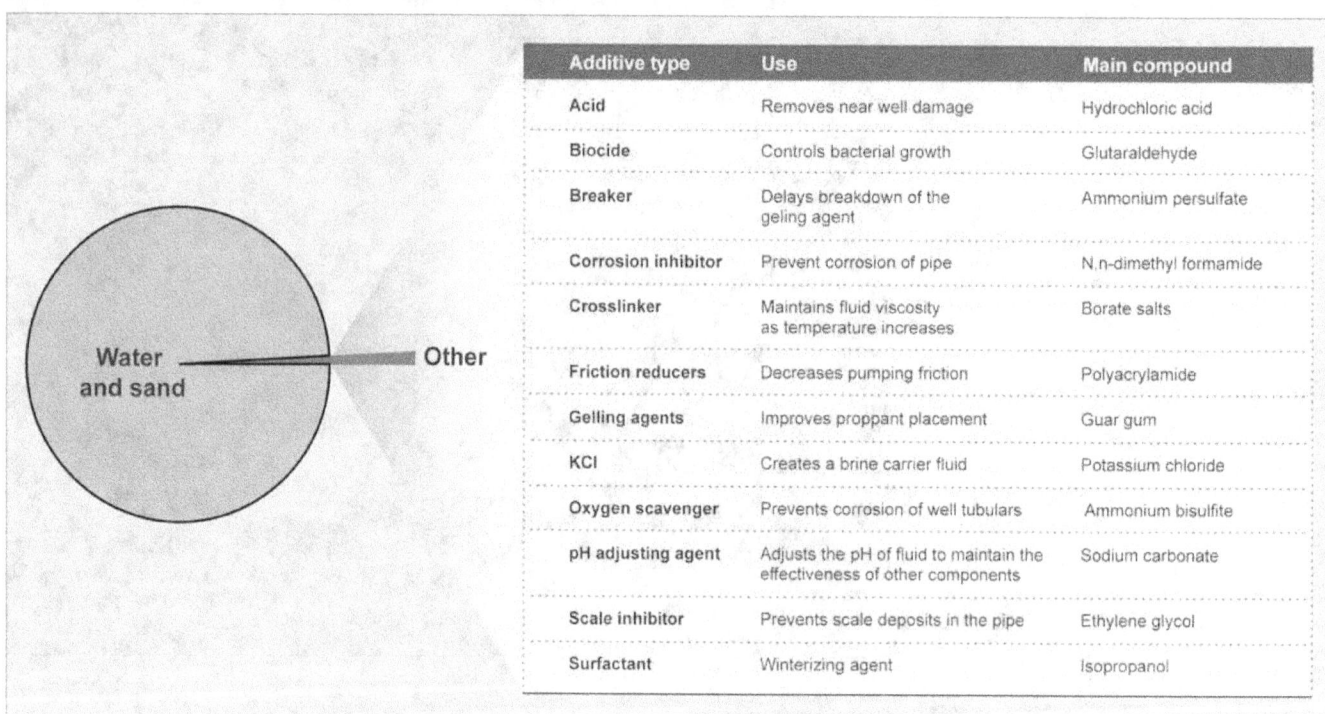

Additive type	Use	Main compound
Acid	Removes near well damage	Hydrochloric acid
Biocide	Controls bacterial growth	Glutaraldehyde
Breaker	Delays breakdown of the geling agent	Ammonium persulfate
Corrosion inhibitor	Prevent corrosion of pipe	N,n-dimethyl formamide
Crosslinker	Maintains fluid viscosity as temperature increases	Borate salts
Friction reducers	Decreases pumping friction	Polyacrylamide
Gelling agents	Improves proppant placement	Guar gum
KCl	Creates a brine carrier fluid	Potassium chloride
Oxygen scavenger	Prevents corrosion of well tubulars	Ammonium bisulfite
pH adjusting agent	Adjusts the pH of fluid to maintain the effectiveness of other components	Sodium carbonate
Scale inhibitor	Prevents scale deposits in the pipe	Ethylene glycol
Surfactant	Winterizing agent	Isopropanol

Sources: Department of Energy and Groundwater Protection Council

The operator pumps the fracturing fluid into the wellbore at pressures high enough to force the fluid through the perforations into the surrounding formation—which can be shale, coalbeds, or tight sandstone—expanding existing fractures and creating new ones in the process. After the fractures are created, the operator reduces the pressure. The proppant stays in the formation to hold open the fractures and allow the release of oil and gas. Some of the fracturing fluid that was injected into the well will return to the surface (commonly referred to as flowback) along with water that occurs naturally in the oil- or gas-bearing formation—collectively referred to as produced water. The produced water is brought to the surface and collected by the operator, where it can be stored on-site in impoundments, injected into underground wells, transported to a wastewater treatment plant, or reused by the operator in other ways.[21] Given the length of horizontal wells, hydraulic fracturing is often conducted in stages, where each stage focuses on a limited linear section and may be repeated numerous times.

Once a well is producing oil or natural gas, equipment and temporary infrastructure associated with drilling and hydraulic fracturing operations is no longer needed and may be removed, leaving only the parts of the infrastructure required to collect and process the oil or gas and ongoing produced water. Operators may begin to reclaim the part of the site that will not be used by restoring the area to predevelopment conditions. Throughout the producing life of an oil or gas well, the operator may find it necessary to periodically restimulate the flow of oil or gas by repeating the hydraulic fracturing process. The frequency of such activity depends on the characteristics of the geologic formation and the economics of the individual well. If the hydraulic fracturing process is repeated, the site and surrounding area will be further affected by the required infrastructure, truck transport, and other activity associated with this process.

Regulatory Framework

Shale oil and gas development, like conventional onshore oil and gas production, is governed by a framework of federal, state, and local laws and regulations. Most shale development in the near future is expected to occur on nonfederal lands and, therefore, states will typically take the lead in regulatory activities. However, in some cases, federal agencies oversee shale oil and gas development. For example, BLM oversees shale oil and gas development on federal lands. In large part, the federal laws, regulations, and permit requirements that apply to conventional onshore oil and gas exploration and production activities also apply to shale oil and gas development.

- *Federal.* A number of federal agencies administer laws and regulations that apply to various phases of shale oil and gas development. For example, BLM manages federal lands and approximately 700 million acres of federal subsurface minerals, also known as the federal mineral estate. EPA administers and enforces key federal laws, such as the Safe Drinking Water Act, to protect

---

[21]Underground injection is the predominant practice for disposing of produced water. In addition to underground injection, a limited amount of produced water is managed by discharging it to surface water, storing it in surface impoundments, and reusing it for irrigation or hydraulic fracturing.

human health and the environment. Other federal land management agencies, such as the U.S. Department of Agriculture's Forest Service and the Department of the Interior's Fish and Wildlife Service, also manage federal lands, including shale oil and gas development on those lands.

- *State.* State agencies implement and enforce many of the federal environmental regulations and may also have their own set of state laws covering shale oil and gas development.

- *Other.* Additional requirements regarding shale oil and gas operations may be imposed by various levels of government for specific locations. Entities such as cities, counties, tribes, and regional water authorities may set additional requirements that affect the location and operation of wells.

GAO is conducting a separate and more detailed review of the federal and state laws and regulations that apply to unconventional oil and gas development, including shale oil and gas.

Location of Shale Oil and Gas in the United States

Shale oil and gas are found in shale plays—a set of discovered or undiscovered oil and natural gas accumulations or prospects that exhibit similar geological characteristics—on private, state-owned, and federal lands across the United States. Shale plays are located within basins, which are large-scale geological depressions, often hundreds of miles across, that also may contain other oil and gas resources. Figure 4 shows the location of shale plays and basins in the contiguous 48 states.

Figure 4: Shale Plays and Basins in the Contiguous 48 States

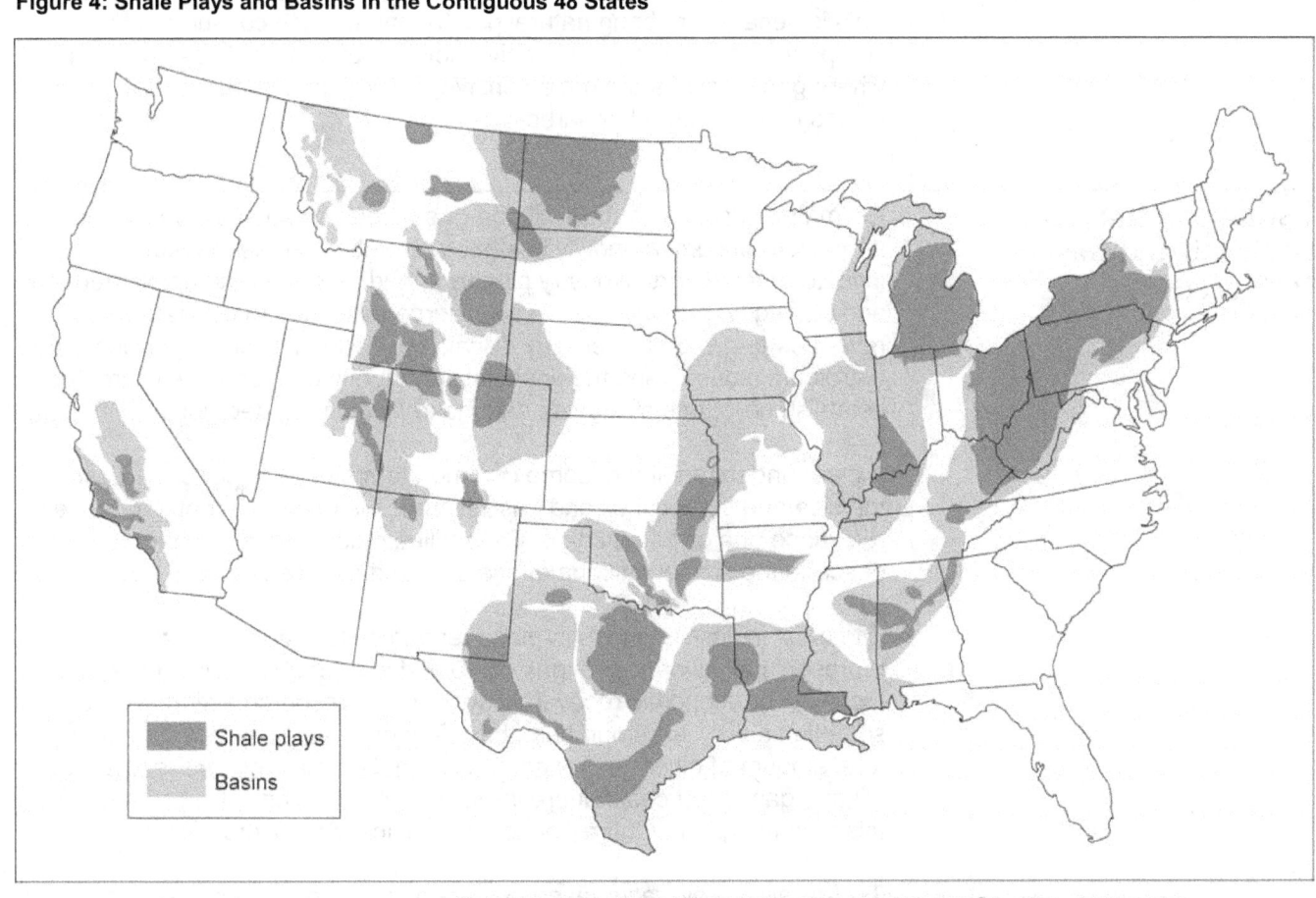

Shale plays

Basins

A shale play can be developed for oil, natural gas, or both. In addition, a shale gas play may contain "dry" or "wet" natural gas. Dry natural gas is a mixture of hydrocarbon compounds that exists as a gas both underground in the reservoir and during production under standard temperature and pressure conditions. Wet natural gas contains natural gas liquids, or the portion of the hydrocarbon resource that exists as a gas when in natural underground reservoir conditions but that is liquid at surface conditions. The natural gas liquids are typically propane, butane, and ethane and are separated from the produced gas at the surface in lease separators, field facilities, or gas processing plants. Operators may then sell the natural gas liquids, which may give wet shale gas plays an economic advantage over dry gas plays. Another advantage of liquid petroleum and natural gas liquids is that they can be transported more easily than natural gas. This is because, to bring natural gas to markets and consumers, companies must build an extensive network of gas pipelines. In areas where gas pipelines are not extensive, natural gas produced along with liquids is often vented or flared.

Estimating the Size of Shale Oil and Gas Resources

Estimating the size of shale oil and gas resources serves a variety of needs for consumers, policymakers, land and resource managers, investors, regulators, industry planners, and others. For example, federal and state governments may use resource estimates to estimate future revenues and establish energy, fiscal, and national security policies. The petroleum industry and the financial community use resource estimates to establish corporate strategies and make investment decisions.

A clear understanding of some common terms used to generally describe the size and scope of oil and gas resources is needed to determine the relevance of a given estimate. For an illustration of how such terms describe the size and scope of shale oil and gas, see figure 5.

The most inclusive term is in-place resource. The in-place resource represents all oil or natural gas contained in a formation without regard to technical or economic recoverability. In-place resource estimates are sometimes very large numbers, but often only a small proportion of the total amount of oil or natural gas in a formation may ever be recovered. Oil and gas resources that are in-place, but not technically recoverable at this time may, in the future, become technically recoverable.

Technically recoverable resources are a subset of in-place resources that include oil or gas, including shale oil and gas that is producible given available technology. Technically recoverable resources include those that are economically producible and those that are not. Estimates of technically recoverable resources are dynamic, changing to reflect the potential of extraction technology and knowledge about the geology and composition of geologic formations. According to the National Petroleum Council,[22] technically recoverable resource estimates usually increase

---

[22]The National Petroleum Council is a federally chartered and privately funded advisory committee that advises, informs, and makes recommendations to the Secretary of Energy on oil and natural gas matters.

over time because of the availability of more and better data, or knowledge of how to develop a new play type (such as shale formations).

Proved reserve estimates are more precise than technically recoverable resources and represent the amount of oil and gas that have been discovered and defined, typically by drilling wells or other exploratory measures, and which can be economically recovered within a relatively short time frame. Proved reserves may be thought of as the "inventory" that operators hold and define the quantity of oil and gas that operators estimate can be recovered under current economic conditions, operating methods, and government regulations. Estimates of proved reserves increase as oil and gas companies make new discoveries and report them to the government; oil and gas companies can increase their reserves as they develop already-discovered fields and improve production technology. Reserves decline as oil and gas reserves are produced and sold. In addition, reserves can change as prices and technologies change. For example, technology improvements that enable operators to extract more oil or gas from existing fields can increase proved reserves. Likewise, higher prices for oil and gas may increase the amount of proved reserves because more resources become financially viable to extract.[23] Conversely, lower prices may diminish the amount of resources likely to be produced, reducing proved reserves.

Historical production refers to the total amount of oil and gas that has been produced up to the present. Because these volumes of oil and gas have been measured historically, this is the most precise information available as it represents actual production amounts.

---

[23]For example, secondary recovery operations can be costly (such as using a well to inject water into an oil reservoir and push any remaining oil to operating wells), but the costs may be justified if prices are high enough.

Figure 5: Common Terminology to Describe the Size and Scope of Shale Oil and Gas

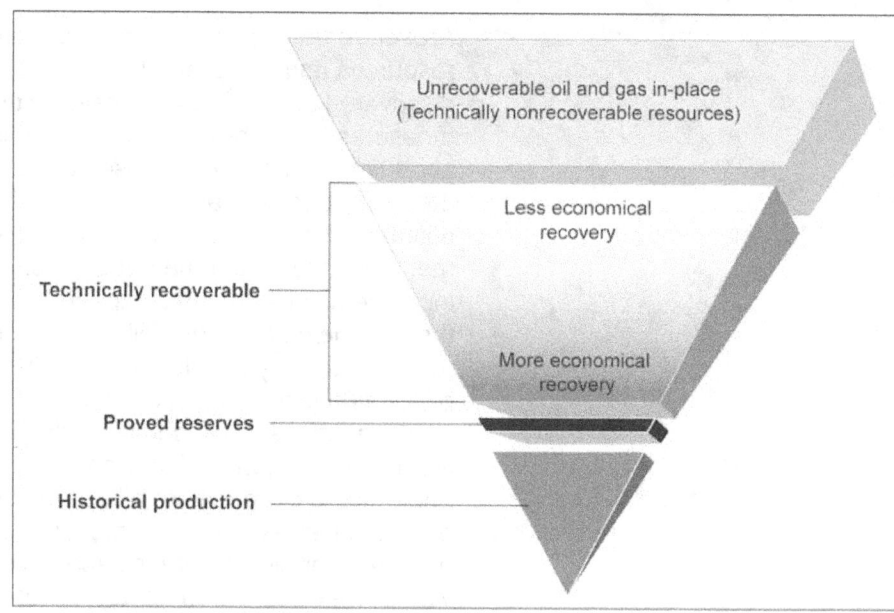

Sources: GAO; based on illustration by the Congressional Research Service.

Note: This illustration is not necessarily to scale because all volumes, except historical production, are subject to significant uncertainty.

Certain federal agencies have statutory responsibility for collecting and publishing authoritative statistical information on various types of energy sources in the United States. EIA collects, analyzes, and disseminates independent and impartial energy information, including data on shale oil and gas resources. Under the Energy Policy and Conservation Act of 2000, as amended, USGS estimates onshore undiscovered technically recoverable oil and gas resources in the United States.[24] USGS has conducted a number of national estimates of undiscovered technically recoverable oil and natural gas resources over several decades. USGS geologists and other experts estimate undiscovered oil and gas—that is, oil and gas that has not been proven to be present by oil and gas companies—based on geological survey data and other information about

---

[24]Pub. L. No. 106-469 § 604 (2000), 114 Stat. 2029, 2041-42, codified, as amended, at 42 U.S.C. § 6217.

the location and size of different geological formations across the United States. In addition to EIA and USGS, experts from industry, academia, federal advisory committees, private consulting firms, and professional societies also estimate the size of the resource.

Domestic Shale Oil and Gas Estimates and Production

Estimates of the size of shale oil and gas resources in the United States have increased over time as has the amount of such resources produced from 2007 through 2011. Specifically, over the last 5 years, estimates of (1) technically recoverable shale oil and gas and (2) proved reserves of shale oil and gas have increased, as technology has advanced and more shale has been drilled. In addition, domestic shale oil and gas production has experienced substantial growth in recent years.

Estimates of Technically Recoverable Shale Oil and Gas Resources

EIA, USGS, and the Potential Gas Committee have increased their estimates of the amount of technically recoverable shale oil and gas over the last 5 years, which could mean an increase in the nation's energy portfolio; however, less is known about the amount of technically recoverable shale oil than shale gas, in part because large-scale production of shale oil has been under way for only the past few years. The estimates are from different organizations and vary somewhat because they were developed at different times and using different data, methods, and assumptions, but estimates from all of these organizations have increased over time, indicating that the nation's shale oil and gas resources may be substantial. For example, according to estimates and reports we reviewed, assuming current consumption levels without consideration of a specific market price for future gas supplies, the amount of domestic technically recoverable shale gas could provide enough natural gas to supply the nation for the next 14 to 100 years. The increases in estimates can largely be attributed to improved geological information about the resources, greater understanding of production levels, and technological advancements.

Estimates of Technically Recoverable Shale Oil Resources

In the last 2 years, EIA and USGS provided estimates of technically recoverable shale oil.[25] Each of these estimates increased in recent years as follows:

- In 2012, EIA estimated that the United States possesses 33 billion barrels of technically recoverable shale oil,[26] mostly located in four shale formations—the Bakken in Montana and North Dakota; Eagle Ford in Texas; Niobrara in Colorado, Kansas, Nebraska, and Wyoming; and the Monterey in California.

- In 2011, USGS estimated that the United States possesses just over 7 billion barrels of technically recoverable oil in shale and tight sandstone formations. The estimate represents a more than threefold increase from the agency's estimate in 2006. However, there are several shale plays that USGS has not evaluated for shale oil because interest in these plays is relatively new. According to USGS officials, these shale plays have shown potential for production in recent years and may contain additional shale oil resources. Table 1 shows USGS' 2006 and 2011 estimates and EIA's 2011 and 2012 estimates.

Table 1: USGS and EIA Estimates of Total Remaining Technically Recoverable U.S. Oil Resources

Barrels of oil in billions

	USGS		EIA	
	2006	2011	2011	2012
Estimated technically recoverable shale oil and tight sandstone resources	2	7	32	33
Estimated technically recoverable oil resources other than shale[a]	142	133	187	201

Source: GAO analysis of EIA and USGS data.

[25]As noted previously, for the purposes of this report, we use the term "shale oil" to refer to oil from shale and other tight formations, which is recoverable by hydraulic fracturing and horizontal drilling techniques and is described by others as "tight oil." Shale oil and tight oil are extracted in the same way, but differ from "oil shale." Oil shale is a sedimentary rock containing solid organic material that converts into a type of crude oil only when heated.

[26]Comparatively, the United States currently consumes about 7 billion barrels of oil per year, about half of which are imported from foreign sources.

aIncludes estimates for conventional offshore oil and gas, as well as natural gas liquids. In addition, the USGS estimates for 2006 and 2011 include a 2006 estimate of technically recoverable offshore conventional oil resources totaling 86 billion barrels of oil and natural gas liquids from the former Minerals Management Service, which has since been reorganized into the Bureau of Ocean Energy Management and the Bureau of Safety and Environmental Enforcement.

Overall, estimates of the size of technically recoverable shale oil resources in the United States are imperfect and highly dependent on the data, methodologies, model structures, and assumptions used. As these estimates are based on data available at a given point in time, they may change as additional information becomes available. Also these estimates depend on historical production data as a key component for modeling future supply. Because large-scale production of oil in shale formations is a relatively recent activity, their long-term productivity is largely unknown. For example, EIA estimated that the Monterey Shale in California may possess about 15.4 billion barrels of technically recoverable oil. However, without a longer history of production, the estimate has greater uncertainty than estimates based on more historical production data. At this time, USGS has not yet evaluated the Monterey Shale play.

Estimates of Technically Recoverable Shale Gas Resources

The amount of technically recoverable shale gas resources in the United States has been estimated by a number of organizations, including EIA, USGS, and the Potential Gas Committee (see fig. 6). Their estimates were as follows:

- In 2012, EIA estimated the amount of technically recoverable shale gas in the United States at 482 trillion cubic feet.[27] This represents an increase of 280 percent from EIA's 2008 estimate.

- In 2011, USGS reported that the total of its estimates for the shale formations the agency evaluated in all previous years[28] shows the

---

[27] EIA estimates are based on natural gas production data from 2 years prior to the reporting year; for example, EIA's 2012 estimate is based on 2010 data; the date cited here reflects the fact that EIA reported this latest estimate in 2012.

[28] USGS estimates are based on updated data in a few—but not all—individual geological areas, combined with data from other areas from all previous years. Each year USGS estimates new information for a few individual geological areas. For example, the 2011 USGS estimate includes updated 2011 data for the Appalachian Basin, the Anadarko Basin, and the Gulf Coast, combined with estimates for all other areas developed before 2011. See appendix III for additional information on USGS estimates. The date cited here reflects the fact that USGS reported this latest estimate in 2011.

amount of technically recoverable shale gas in the United States at about 336 trillion cubic feet. This represents an increase of about 600 percent from the agency's 2006 estimate.

- In 2011, the Potential Gas Committee estimated the amount of technically recoverable shale gas in the United States at about 687 trillion cubic feet.[29] This represents an increase of 240 percent from the committee's 2007 estimate.

Figure 6: Estimates of Technically Recoverable Shale Gas from EIA, USGS, and the Potential Gas Committee (2006 through 2012)

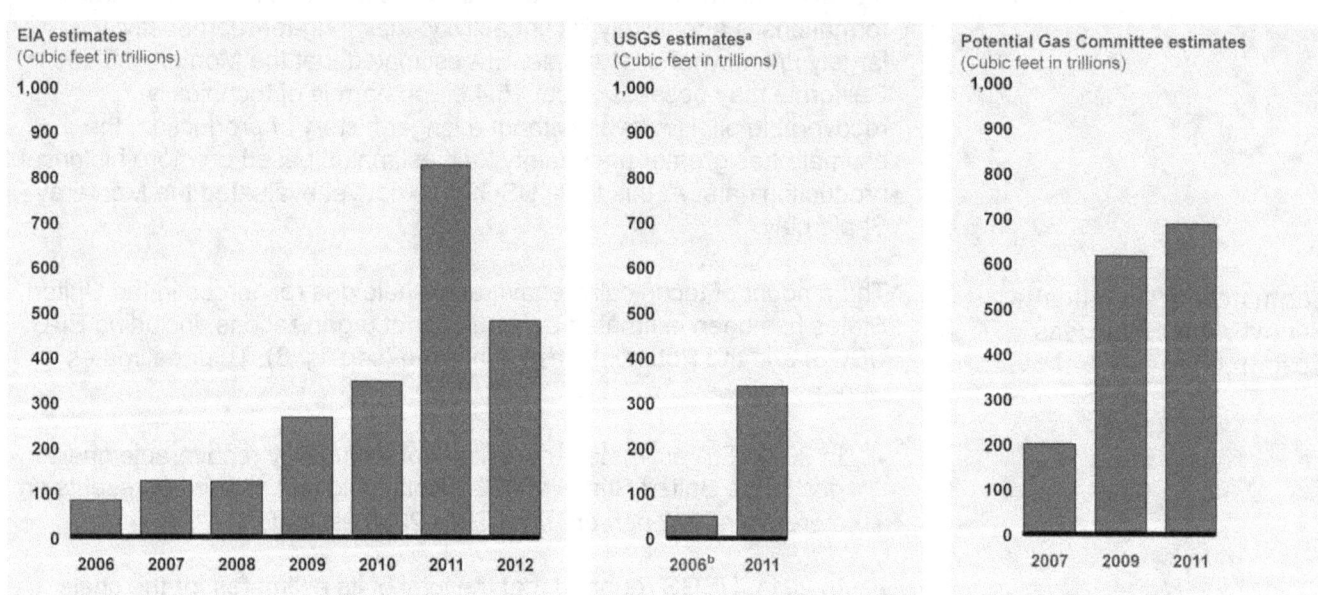

Sources: GAO analysis of EIA, Potential Gas Commitee, and USGS estimates.

Notes: Natural gas is generally priced and sold in thousand cubic feet (abbreviated Mcf, using the Roman numeral for 1,000). Units of a trillion cubic feet (Tcf) are often used to measure large quantities, as in resources or reserves in the ground, or annual national energy consumption. One Tcf is enough natural gas to heat 15 million homes for 1 year or fuel 12 million natural gas-fired vehicles for 1 year. In 2012, EIA reduced its estimate of technically recoverable shale gas in the Marcellus Shale by about 67 percent. According to EIA officials, the decision to revise the estimate was based primarily on the availability of new production data, which was highlighted by the release of the USGS

---

[29]Potential Gas Committee estimates are based on natural gas production data from the previous year; for example, committee's 2011 estimate is based on 2010 data. The date cited here reflects the fact that the Potential Gas Committee reported this latest estimate in 2011.

estimate. In 2011, EIA used data from a contractor to estimate that the Marcellus Shale possessed about 410 trillion cubic feet of technically recoverable gas. After EIA released its estimates in 2011, USGS released its first estimate of technically recoverable gas in the Marcellus in almost 10 years. USGS estimated that there were 84 trillion cubic feet of natural gas in the Marcellus—which was 40 times more than its previous estimate reported in 2002 but significantly less than EIA's estimate. In 2012, EIA announced that it was revising its estimate of the technically recoverable gas in the Marcellus Shale from 410 to 141 trillion cubic feet. EIA reported additional details about its methodology and data in June 2012. See U.S. Department of Energy, Energy Information Administration, Annual Energy Outlook 2012, With Projections to 2035 (DOE/EIA-0383 [2012], Washington, D.C., June 25, 2012).

[a]The 2006 USGS estimate of about 54 trillion cubic feet represents those assessments that had been done up to the end of 2006. As such, the estimate is partially dependent on how the agency scheduled basin studies and assessments from 2000 through 2006, rather than purely on changes in USGS views of resource potential since 2006.

[b]The Potential Gas Committee did not report separate estimates of shale gas until 2007 and has updated this estimate every 2 years since then.

In addition to the estimates from the three organizations we reviewed, operators and energy forecasting consultants prepare their own estimates of technically recoverable shale gas to plan operations or for future investment. In September 2011, the National Petroleum Council aggregated data on shale gas resources from over 130 industry, government, and academic groups and estimated that approximately 1,000 trillion cubic feet of shale gas is available for production domestically. In addition, private firms that supply information to the oil and gas industry conduct assessments of the total amount of technically recoverable natural gas. For example, ICF International, a consulting firm that provides information to public- and private-sector clients, estimated in March 2012 that the United States possesses about 1,960 trillion cubic feet of technically recoverable shale gas.

Based on estimates from EIA, USGS, and the Potential Gas Committee, five shale plays—the Barnett, Haynesville, Fayetteville, Marcellus, and Woodford—are estimated to possess about two-thirds of the total estimated technically recoverable gas in the United States (see table 2).

Table 2: Estimated Technically Recoverable Shale Gas Resources, by Play

Shale play	Location	Technically recoverable gas, in trillion cubic feet (Tcf)
Barnett	North Texas	43-53
Fayetteville	Arkansas	13-110
Haynesville	Louisiana and East Texas	66-110
Marcellus	Northeast United States	84-227[a]
Woodford	Oklahoma	11-27

Sources: GAO analysis of EIA, USGS, and Potential Gas Committee data.

Note: The estimated technically recoverable gas shown here represents the range of estimates for these plays determined by EIA, USGS, and the Potential Gas Committee.

[a]This estimate of the Marcellus also includes estimated shale gas from other nearby lands in the Appalachian area; but, according to an official for the estimating organization, the Marcellus Shale is the predominant source of gas in the basin.

As with estimates for technically recoverable shale oil, estimates of the size of technically recoverable shale gas resources in the United States are also highly dependent on the data, methodologies, model structures, and assumptions used and may change as additional information becomes available. These estimates also depend on historical production data as a key component for modeling future supply. Because most shale gas wells generally were not in place until the last few years, their long-term productivity is untested. According to a February 2012 report released by the Sustainable Investments Institute and the Investor Responsibility Research Center Institute, production in emerging shale plays has been concentrated in areas with the highest known gas production rates, and many shale plays are so large that most of the play has not been extensively tested.[30] As a result, production rates achieved to date may not be representative of future production rates across the formation. EIA reports that experience to date shows production rates from neighboring shale gas wells can vary by as much as a factor of 3 and that production rates for different wells in the same formation can vary by as much as a factor of 10. Most gas companies estimate that production in a given well will drop sharply after the first few years and

---

[30]The Sustainable Investments Institute (Si2) is a nonprofit membership organization founded in 2010 to conduct research and publish reports on organized efforts to influence corporate behavior. The Investor Responsibility Research Center Institute is a nonprofit organization established in 2006 that provides information to investors.

then level off, continuing to produce gas for decades, according to the Sustainable Investments Institute and the Investor Responsibility Research Center Institute.

Estimates of Proved Reserves of Shale Oil and Gas

Estimates of proved reserves of shale oil and gas increased from 2007 to 2009. Operators determine the size of proved reserves based on information collected from drilling, geological and geophysical tests, and historical production trends. These are also the resources operators believe they will develop in the short term—generally within the next 5 years—and assume technological and economic conditions will remain unchanged.

Estimates of proved reserves of shale oil. EIA does not report proved reserves of shale oil separately from other oil reserves; however, EIA and others have noted an increase in the proved reserves of oil in the nation, and federal officials attribute the increase, in part, to oil from shale and tight sandstone formations. For example, EIA reported in 2009 that the Bakken Shale in North Dakota and Montana drove increases in oil reserves, noting that North Dakota proved reserves increased over 80 percent from 2008 through 2009.

Estimates of proved reserves of shale gas. According to data EIA collects from about 1,200 operators, proved reserves of shale gas have grown from 23 trillion cubic feet in 2007 to 61 trillion cubic feet in 2009, or an increase of 160 percent.[31] More than 75 percent of the proved shale gas reserves are located in three shale plays—the Barnett, Fayetteville, and the Haynesville.

Shale Oil and Gas Production

From 2007 through 2011, annual production of shale oil and gas has experienced significant growth. Specifically, shale oil production increased more than fivefold, from 39 to about 217 million barrels over this 5-year period, and shale gas production increased approximately fourfold, from 1.6 to about 7.2 trillion cubic feet, over the same period. To

---

[31]Reserves are key information for assessing the net worth of an operator. Oil and gas companies traded on the U.S. stock exchange are required to report their reserves to the Securities and Exchange Commission. According to an EIA official, EIA reports a more complete measure of oil and gas reserves because it receives reports of proved reserves from both private and publically held companies.

put this shale production into context, the annual domestic consumption of oil in 2011 was about 6,875 million barrels of oil, and the annual consumption of natural gas was about 24 trillion cubic feet. The increased shale oil and gas production was driven primarily by technological advances in horizontal drilling and hydraulic fracturing that made more shale oil and gas development economically viable.

Shale Oil Production

Annual shale oil production in the United States increased more than fivefold, from about 39 million barrels in 2007 to about 217 million barrels in 2011, according to data from EIA (see fig. 7).[32] This is because new technologies allowed more oil to be produced economically, and because of recent increases in the price for liquid petroleum that have led to increased investment in shale oil development.

Figure 7: Estimated Production of Shale Oil from 2007 through 2011 (in millions of barrels of oil)

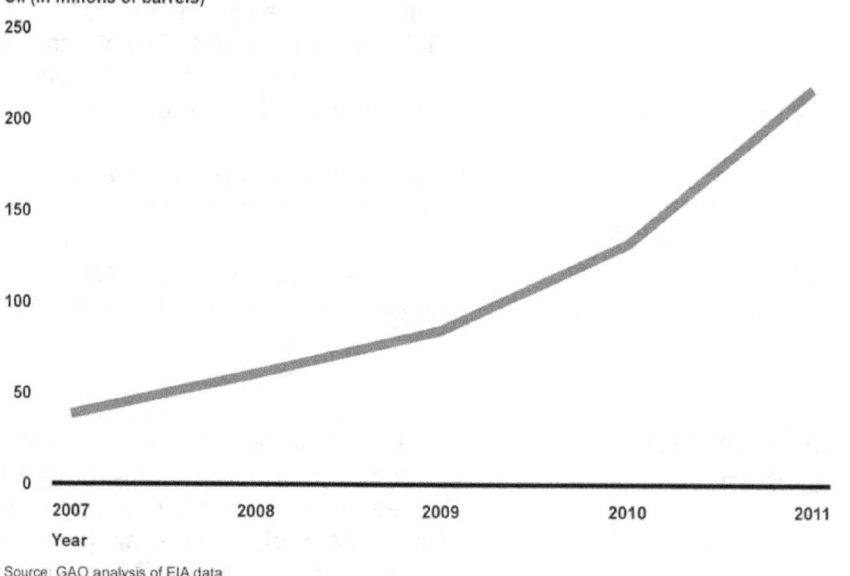

Source: GAO analysis of EIA data.

[32]As noted previously, for the purposes of this report, we use the term "shale oil" to refer to oil from shale and other tight formations, which is recovered by hydraulic fracturing and horizontal drilling and is described by others as "tight oil." Shale oil and tight oil are extracted in the same way, but differ from "oil shale." Oil shale is a sedimentary rock containing solid organic material that converts into a type of crude oil only when heated.

In total, during this period, about 533 million barrels of shale oil was produced. More than 65 percent of the oil was produced in the Bakken Shale (368 million barrels; see fig. 8).[33] The remainder was produced in the Niobrara (62 million barrels), Eagle Ford (68 million barrels), Monterey (18 million barrels), and the Woodford (9 million barrels). To put this in context, shale oil production from these plays in 2011 constituted about 8 percent of U.S. domestic oil consumption, according to EIA data.[34]

---

[33]EIA provided us with estimated shale oil production data from a contractor, HPDI LLC., for 2007 through 2011. EIA uses these data for the purposes of estimating recent shale oil production. EIA has not routinely reported shale oil production data separately from oil production.

[34]In addition to production from these shale oil plays, EIA officials told us that oil was produced from "tight oil" plays such as the Austin Chalk. The technology for producing tight oil is the same as for shale oil, and EIA uses the term "tight oil" to encompass both shale oil and tight oil that are developed with the same type of technology. In addition, EIA officials added that the shale oil data presented here is approximate because the data comes from a sample of similar plays. Overtime, this production data will become more precise as more data becomes available to EIA.

Figure 8: Shale Oil Production, by Shale Play (from 2007 through 2011)

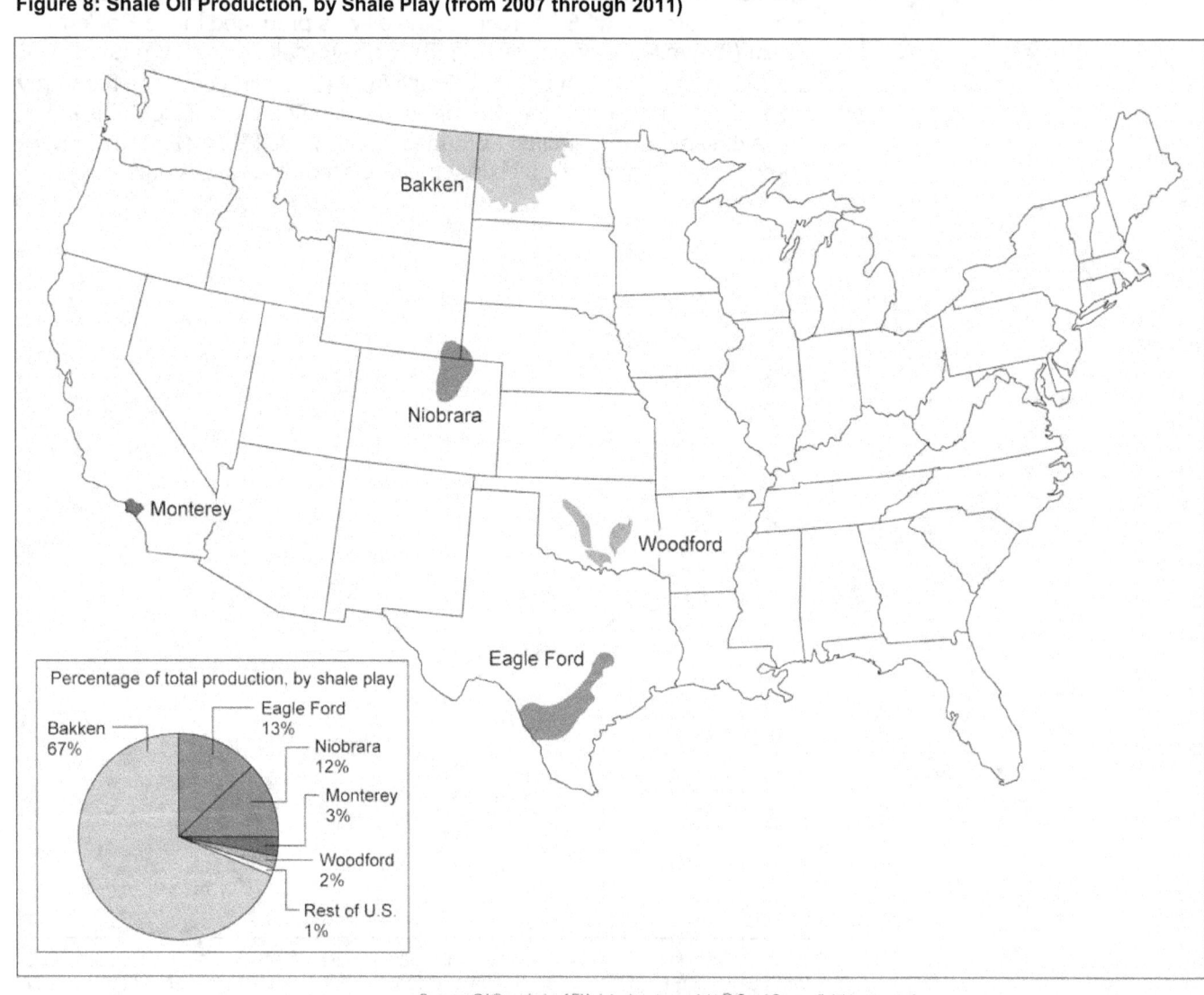

Percentage of total production, by shale play

Bakken 67%
Eagle Ford 13%
Niobrara 12%
Monterey 3%
Woodford 2%
Rest of U.S. 1%

Shale Gas Production

Shale gas production in the United States increased more than fourfold, from about 1.6 trillion cubic feet in 2007 to about 7.2 trillion cubic feet in 2011, according to estimated data from EIA (see fig. 9).[35]

Figure 9: Estimated Production of Shale Gas from 2007 through 2011 (in trillions of cubic feet)

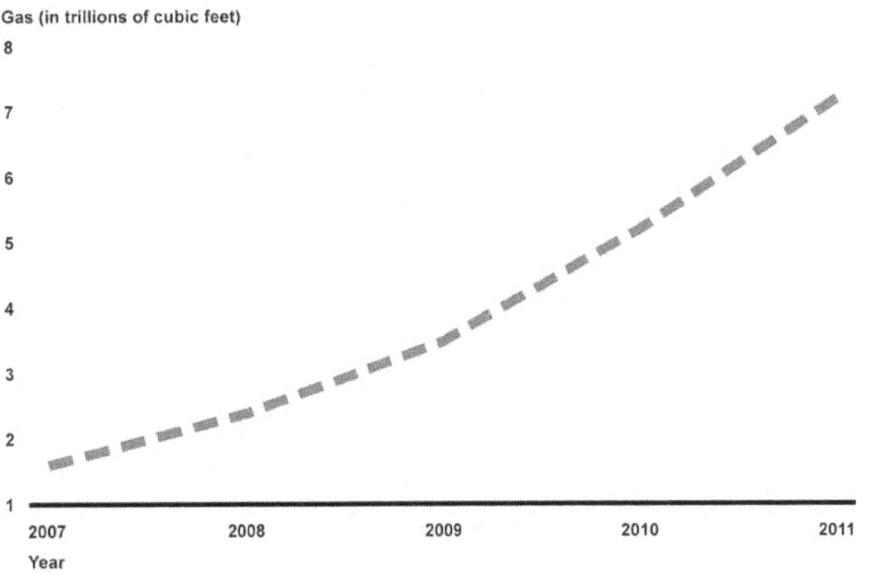

Source: GAO analysis of EIA data.

In total, during this period, about 20 trillion cubic feet of shale gas was produced—representing about 300 days of U.S. consumption, based on 2011 consumption rates. More than 75 percent of the gas was produced in four shale plays—the Barnett, Marcellus, Fayetteville, and Haynesville (see fig.10). From 2007 through 2011, shale gas' contribution to the nation's total natural gas supply grew from about 6 percent in 2007 to approximately 25 percent in 2011 and is projected, under certain assumptions, to increase to 49 percent by 2035, according to an EIA report. Overall production of shale gas increased from calendar years 2007 through 2011, but production of natural gas on federal and tribal

[35]EIA provided us with estimated shale gas production data from a contractor, Lippman Consulting, Inc., for 2007 through 2011. EIA uses these data for the purposes of estimating recent shale gas production. EIA has separately reported shale gas production data using reports from states for the years 2008 and 2009.

lands—including shale gas and natural gas from all other sources—decreased by about 17 percent, according to an EIA report. EIA attributes this decrease to several factors, including the location of shale formations—which, according to an EIA official, appear to be predominately on nonfederal lands.

Figure 10: Shale Gas Production, by Shale Play (from 2007 through 2011)

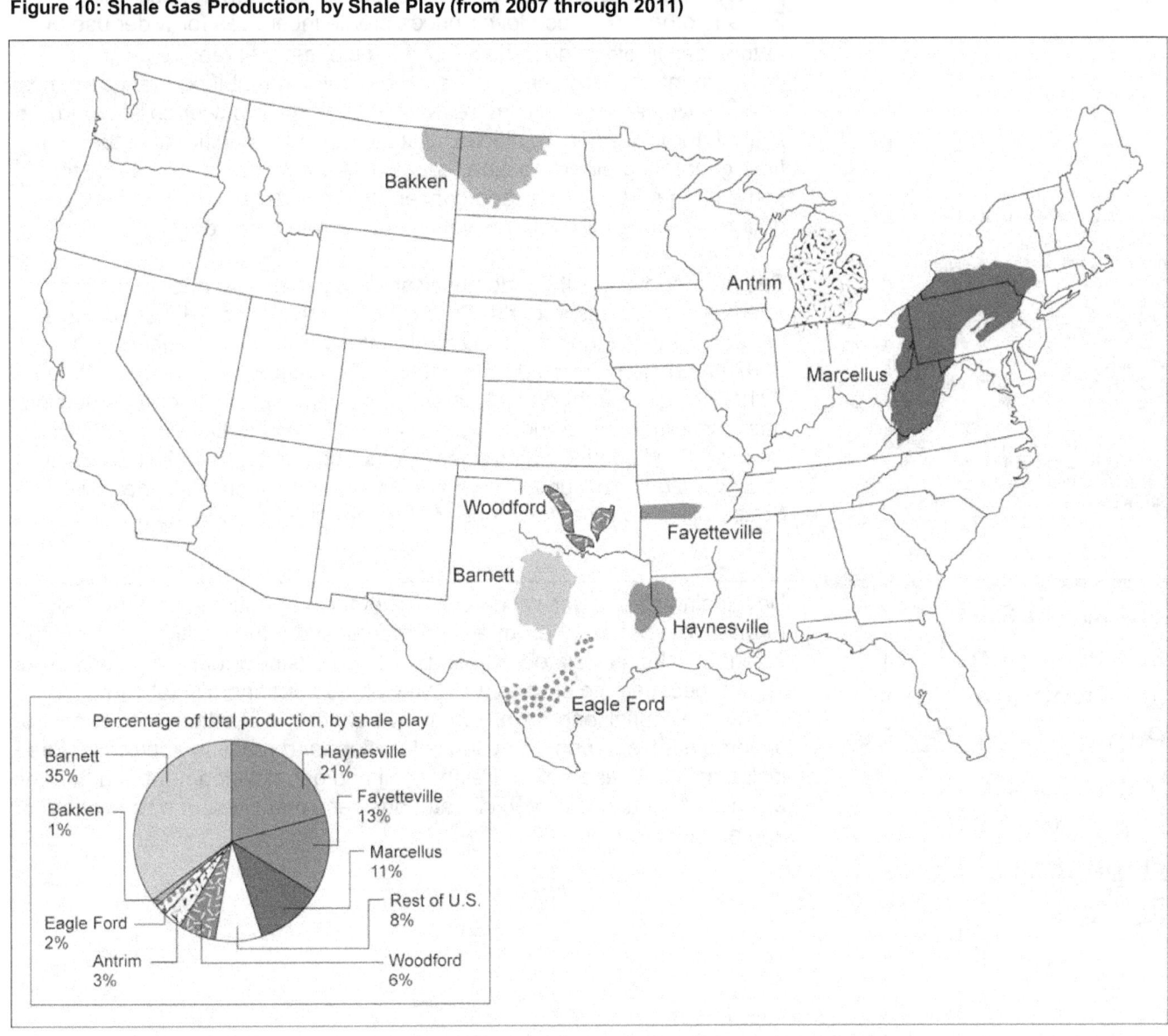

The growth in production of shale gas has increased the overall supply of natural gas in the U.S. energy market. Since 2007, increased shale gas production has contributed to lower prices for consumers, according to EIA and others.[36] These lower prices create incentives for wider use of natural gas in other industries. For example, several reports by government, industry, and others have observed that if natural gas prices remain low, natural gas is more likely to be used to power cars and trucks in the future. In addition, electric utilities may build additional natural gas-fired generating plants as older coal plants are retired. At the same time, some groups have expressed concern that greater reliance on natural gas may reduce interest in developing renewable energy.

The greater availability of domestic shale gas has also decreased the need for natural gas imports. For example, EIA has noted that volumes of natural gas imported into the United States have fallen in recent years—in 2007, the nation imported 16 percent of the natural gas consumed and in 2010, the nation imported 11 percent—as domestic shale gas production has increased. This trend is also illustrated by an increase in applications for exporting liquefied natural gas to other countries. In its 2012 annual energy outlook, EIA predicted that, under certain scenarios, the United States will become a net exporter of natural gas by about 2022.[37]

Development of Wet Gas

EIA reported that operators have recently moved away from the development of shale plays that are primarily dry gas in favor of developing plays with higher concentrations of natural gas liquids. At current natural gas prices, natural gas liquids are a much more valuable product than dry gas. This is because the end products and byproducts of natural gas liquids contain more energy per unit of volume and have uses beyond heating and power generation and may be converted into products that can be more easily transported and traded in the global market. Shale plays with significant natural gas liquids include the Eagle Ford and Marcellus.

Shale Oil and Gas Development Pose Environmental and Public Health Risks, but the Extent is Unknown and Depends on Many Factors

Developing oil and gas resources—whether conventional or from shale formations—poses inherent environmental and public health risks, but the extent of risks associated with shale oil and gas development is unknown, in part, because the studies we reviewed do not generally take into account potential long-term, cumulative effects. In addition, the severity of adverse effects depend on various location- and process-specific factors, including the location of future shale oil and gas development and the rate at which it occurs, geology, climate, business practices, and regulatory and enforcement activities.

---

[36]According to a 2012 report from the Bipartisan Policy Center, natural gas prices declined roughly 37 percent from February 2008 to January 2010.

[37]Department of Energy, Energy Information Administration, *Annual Energy Outlook 2012, With Projections to 2035*, DOE/EIA-0383 (Washington, D.C.: June 25, 2012).

Shale Oil and Gas Development Pose Risks to Air, Water, Land and Wildlife

Air Quality

Oil and gas development, which includes development from shale formations, poses inherent risks to air quality, water quantity, water quality, and land and wildlife.

According to a number of studies and publications we reviewed, shale oil and gas development pose risks to air quality. These risks are generally the result of engine exhaust from increased truck traffic, emissions from diesel-powered pumps used to power equipment, intentional flaring or venting of gas for operational reasons, and unintentional emissions of pollutants from faulty equipment or impoundments.

Construction of the well pad, access road, and other drilling facilities requires substantial truck traffic, which degrades air quality. According to a 2008 National Park Service report, an average well, with multistage fracturing, can require 320 to 1,365 truck loads to transport the water, chemicals, sand, and other equipment—including heavy machinery like bulldozers and graders—needed for drilling and fracturing. The increased traffic creates a risk to air quality as engine exhaust that contains air pollutants such as nitrogen oxides and particulate matter that affect public health and the environment are released into the atmosphere.[38] Air quality may also be degraded as fleets of trucks traveling on newly graded or unpaved roads increase the amount of dust released into the air—which can contribute to the formation of regional haze.[39] In addition to the dust, silica sand (see fig. 11)—commonly used as proppant in the hydraulic fracturing process—may pose a risk to human health, if not properly handled. According to a federal researcher from the Department of Health and Human Services, uncontained sand particles and dust pose threats to workers at hydraulic fracturing well sites. The official stated that particles from the sand, if not properly contained by dust control mechanisms, can lodge in the lungs and potentially cause silicosis.[40]

---

[38]Nitrogen oxides are regulated pollutants commonly known as NOx that, among other things, contribute to the formation of ozone and have been linked to respiratory illness, decreased lung function, and premature death. Particulate matter is a ubiquitous form of air pollution commonly referred to as soot. GAO, *Diesel Pollution: Fragmented Federal Programs That Reduce Mobile Source Emissions Could Be Improved*, GAO-12-261 (Washington, D.C.: Feb. 7, 2012).

[39]T. Colborn, C. Kwiatkowski, K. Schultz, and M. Bachran, "Natural Gas Operations From a Public Health Perspective," *International Journal of Human & Ecological Risk Assessment* 17, no. 5 (2011).

[40]Silicosis is an incurable lung disease caused by inhaling fine dusts of silica sand.

The researcher expects to publish the results of research on public health risks from proppant later in 2012.

Figure 11: Silica Sand Proppant

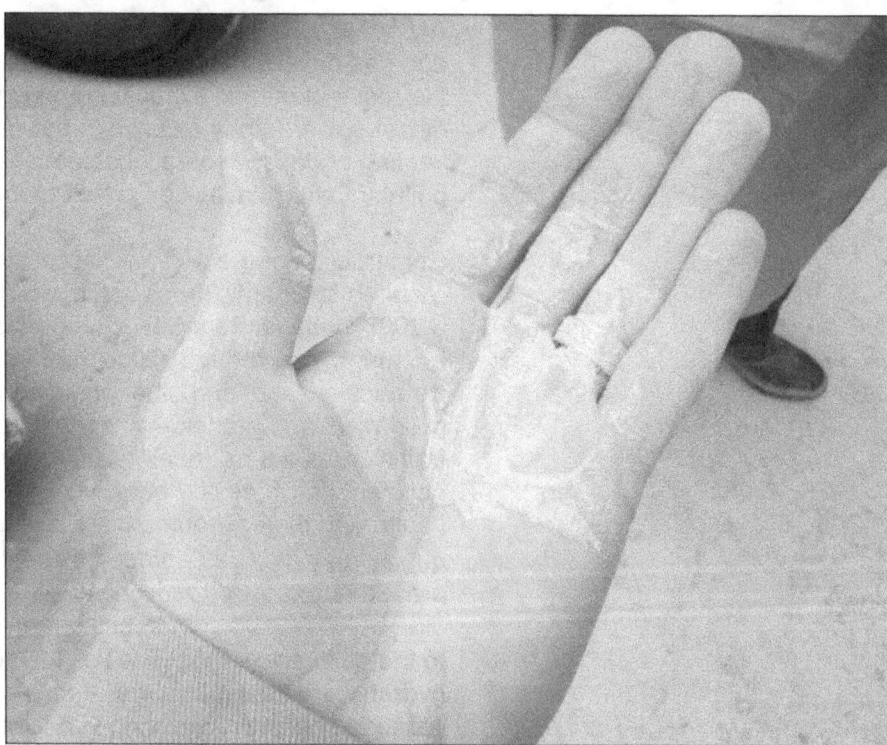

Source: GAO.

Use of diesel engines to supply power to drilling sites also degrades air quality. Shale oil and gas drilling rigs require substantial power to drill and case wellbores to the depths of shale formations. This power is typically provided by transportable diesel engines, which generate exhaust from the burning of diesel fuel. After the wellbore is drilled to the target formation, additional power is needed to operate the pumps that move large quantities of water, sand, or chemicals into the target formation at high pressure to hydraulically fracture the shale—generating additional exhaust. In addition, other equipment used during operations—including pneumatic valves and dehydrators—contribute to air emissions. For example, natural gas powers switches that turn valves on and off in the production system. Each time a valve turns on or off, it "bleeds" a small amount of gas into the air. Some of these pneumatic valves vent gas continuously. A dehydrator circulates the chemical glycol to absorb moisture in the gas but also absorbs small volumes of gas. The absorbed gas vents to the atmosphere when the water vapor is released from the glycol.[41]

Releases of natural gas during the development process also degrade air quality. As part of the process to develop shale oil and gas resources, operators flare or vent natural gas for a number of operational reasons, including lowering the pressure to ensure safety or when operators purge water or hydrocarbon liquids that collect in wellbores to maintain proper well function. Flaring emits carbon dioxide, and venting releases methane and volatile organic compounds. Venting and flaring are often a necessary part of the development process but contribute to greenhouse gas emissions.[42] According to EPA analysis, natural gas well completions involving hydraulic fracturing vent approximately 230 times more natural gas and volatile organic compounds than natural gas well completions that do not involve hydraulic fracturing.[43] As we reported in July 2004, in addition to the operational reasons for flaring and venting, in areas where the primary purpose of drilling is to produce oil, operators flare or vent associated natural gas because no local market exists for the gas and transporting to a market may not be economically feasible.[44] For example, according to EIA, in 2011, approximately 30 percent of North Dakota's natural gas production from the Bakken Shale was flared by operators due to insufficient natural gas gathering pipelines, processing plants, and transporting pipelines. The percentage of flared gas in North Dakota is considerably higher than the national average; EIA reported that, in 2009,

---

[41]GAO-11-34.

[42]Methane and other chemical compounds found in the earth's atmosphere create a greenhouse effect. Under normal conditions, when sunlight strikes the earth's surface, some of it is reflected back toward space as infrared radiation or heat. Greenhouse gases such as carbon dioxide and methane impede this reflection by trapping heat in the atmosphere. While these gases occur naturally on earth and are emitted into the atmosphere, the expanded industrialization of the world over the last 150 years has increased the amount of emissions from human activity (known as anthropogenic emissions) beyond the level that the earth's natural processes can handle.

[43]EPA, Regulatory Impact Analysis: Final New Source Performance Standards and Amendments to the National Emissions Standards for Hazardous Air Pollutants for the Oil and Natural Gas industry (Research Triangle Park, NC: April 2012).

[44]GAO, *Natural Gas Flaring and Venting: Opportunities to Improve Data and Reduce Emissions*, GAO-04-809 (Washington, D.C.: July 14, 2004).

less than 1 percent of natural gas produced in the United States was vented or flared.

Storing fracturing fluid and produced water in impoundments may also pose a risk to air quality as evaporation of the fluids have the potential to release contaminants into the atmosphere. According to the New York Department of Environmental Conservation's 2011 Supplemental Generic Environmental Impact Statement, analysis of air emission rates of some of the compounds used in the fracturing fluids in the Marcellus Shale reveals the potential for emissions of hazardous air pollutants, in particular methanol, from the fluids stored in impoundments.

As with conventional oil and gas development, emissions can also occur as faulty equipment or accidents, such as leaks or blowouts, release concentrations of methane and other gases into the atmosphere. For example, corrosion in pipelines or improperly tightened valves or seals can be sources of emissions. In addition, according to EPA officials, storage vessels for crude oil, condensate, or produced water are significant sources of methane, volatile organic compounds and hazardous air pollutant emissions.

A number of studies we reviewed evaluated air quality at shale gas development sites. However, these studies are generally anecdotal, short-term, and focused on a particular site or geographic location. For example, in 2010, the Pennsylvania Department of Environmental Protection conducted short-term sampling of ambient air concentrations in north central Pennsylvania. The sampling detected concentrations of natural gas constituents including methane, ethane, propane, and butane in the air near Marcellus Shale drilling operations, but according to this state agency, the concentration levels were not considered significant enough to cause adverse health effects.[45]

The studies and publications we reviewed provide information on air quality conditions at a specific site at a specific time but do not provide the information needed to determine the overall cumulative effect that

---

[45]Methane emissions represent a waste of resources and a fractional contribution to greenhouse gas levels.

shale oil and gas activities have on air quality.[46] The cumulative effect shale oil and gas activities have on air quality will be largely determined by the amount of development and the rate at which it occurs, and the ability to measure this will depend on the availability of accurate information on emission levels. However, the number of wells that will ultimately be drilled cannot be known in advance—in part because the productivity of any particular formation at any given location and depth is not known until drilling occurs. In addition, as we reported in 2010, data on the severity or amount of pollutants released by oil and gas development, including the amount of fugitive emissions, are limited.

Water Quantity

According to a number of studies and publications we reviewed, shale oil and gas development poses a risk to surface water and groundwater because withdrawing water from streams, lakes, and aquifers for drilling and hydraulic fracturing could adversely affect water sources.[47] Operators use water for drilling, where a mixture of clay and water (drilling mud) is used to carry rock cuttings to the surface, as well as to cool and lubricate the drill bit. Water is also the primary component of fracturing fluid. Table 3 shows the average amount of freshwater used to drill and fracture a shale oil or gas well.

Table 3: Average Freshwater Use per Well for Drilling and Hydraulic Fracturing

Shale play	Average freshwater used (in gallons)	
	For drilling	For hydraulic fracturing
Barnett	250,000	4,600,000
Eagle Ford	125,000	5,000,000
Haynesville	600,000	5,000,000
Marcellus	85,000	5,600,000
Niobrara	300,000	3,000,000

Source: GAO analysis of data reported by George King, Apache Corporation (2011).

Note: The amount of water required to hydraulically fracture a single well varies considerably as fracturing of shale oil and gas becomes dominated by more complex, multistaged fracturing activities.

---

[46]According to a 2008 National Park Service report, on a site-by-site basis, emissions may not be significant but on a regional basis may prove significant as states and parks manage regional ozone transport.

[47]An aquifer is an underground layer of rock or unconsolidated sand, gravel, or silt that will yield groundwater to a well or spring.

GAO-12-732 Shale Oil and Gas Development

According to a 2012 University of Texas study,[48] water for these activities is likely to come from surface water (rivers, lakes, ponds), groundwater aquifers, municipal supplies, reused wastewater from industry or water treatment plants, and recycling water from earlier fracturing operations.[49] As we reported in October 2010, withdrawing water from nearby streams and rivers could decrease flows downstream, making the streams and rivers more susceptible to temperature changes—increases in the summer and decreases in the winter. Elevated temperatures could adversely affect aquatic life because many fish and invertebrates need specific temperatures for reproduction and proper development. Further, decreased flows could damage or destroy riparian vegetation. Similarly, withdrawing water from shallow aquifers—an alternative water source—could temporarily affect groundwater resources. Withdrawals could lower water levels within these shallow aquifers and the nearby streams and springs to which they are connected. Extensive withdrawals could reduce groundwater discharge to connected streams and springs, which in turn could damage or remove riparian vegetation and aquatic life. Withdrawing water from deeper aquifers could have longer-term effects on groundwater and connected streams and springs because replenishing deeper aquifers with precipitation generally takes longer.[50] Further, groundwater withdrawal could affect the amount of water available for other uses, including public and private water supplies.

Freshwater is a limited resource in some arid and semiarid regions of the country where an expanding population is placing additional demands on water. The potential demand for water is further complicated by years of drought in some parts of the country and projections of a warming climate. According to a 2011 Massachusetts Institute of Technology study,[51] the amount of water used for shale gas development is small in

---

[48]Charles G. Groat, Ph.D. and Thomas W. Grimshaw, Ph.D., *Fact-Based Regulation for Environmental Protection in Shale Gas Development* (Austin, Texas: The Energy Institute, The University of Texas at Austin, February, 2012).

[49]Operators are pursuing a variety of techniques and technologies to reduce freshwater demand, such as recycling their own produced water and hydraulic fracturing fluids. We recently reported that some shale gas operators have begun reusing produced water for hydraulic fracturing of additional wells (see GAO-12-156).

[50]GAO-11-35.

[51]Massachusetts Institute of Technology, *The Future of Natural Gas: An Interdisciplinary MIT Study* (2011) (web.mit.edu/mitei/research/studies/report-natural-gas.pdf).

comparison to other water uses, such as agriculture and other industrial purposes. However, the cumulative effects of using surface water or groundwater at multiple oil and gas development sites can be significant at the local level, particularly in areas experiencing drought conditions.

Similar to shale oil and gas development, development of gas from coalbed methane formations poses a risk of aquifer depletion. To develop natural gas from such formations, water from the coal bed is withdrawn to lower the reservoir pressure and allow the methane to desorb from the coal. According to a 2001 USGS report, dewatering coalbed methane formations in the Powder River Basin in Wyoming can lower the groundwater table and reduce water available for other uses, such as livestock and irrigation.[52]

The key issue for water quantity is whether the total amount of water consumed for the development of shale oil and gas will result in a significant long-term loss of water resources within a region, according to a 2012 University of Texas study. This is because water used in shale oil and gas development is largely a consumptive use and can be permanently removed from the hydrologic cycle, according to EPA and Interior officials. However, it is difficult to determine the long-term effect on water resources because the scale and location of future shale oil and gas development operations remains largely uncertain. Similarly, the total volume that operators will withdraw from surface water and aquifers for drilling and hydraulic fracturing is not known until operators submit applications to the appropriate regulatory agency. As a result, the cumulative amount of water consumed over the lifetime of the activity— key information needed to assess the effects of water withdrawals— remains largely unknown.

Water Quality

According to a number of studies and publications we reviewed, shale oil and gas development pose risks to water quality from contamination of surface water and groundwater as a result of spills and releases of produced water, chemicals, and drill cuttings; erosion from ground disturbances; or underground migration of gases and chemicals.

---

[52]USGS, *A Field Conference On Impacts of Coalbed Methane Development in the Powder River Basin, Wyoming*, Open-File Report 01-126 (Denver, CO: 2001).

Spills and Releases

Shale oil and gas development poses a risk to water quality from spills or releases of toxic chemicals and waste that can occur as a result of tank ruptures, blowouts, equipment or impoundment failures, overfills, vandalism, accidents (including vehicle collisions), ground fires, or operational errors. For example, tanks storing toxic chemicals or hoses and pipes used to convey wastes to the tanks could leak, or impoundments containing wastes could overflow as a result of extensive rainfall. According to New York Department of Environmental Conservation's 2011 Supplemental Generic Environmental Impact Statement, spilled, leaked, or released chemicals or wastes could flow to a surface water body or infiltrate the ground, reaching and contaminating subsurface soils and aquifers. In August 2003, we reported that damage from oil and gas related spills on National Wildlife Refuges varied widely in severity, ranging from infrequent small spills with no known effect on wildlife to large spills causing wildlife death and long-term water and soil contamination.[53]

Drill cuttings, if improperly managed, also pose a risk to water quality. Drill cuttings brought to the surface during oil and gas development may contain naturally occurring radioactive materials (NORM),[54] along with other decay elements (radium-226 and radium-228), according to an industry report presented at the Society of Petroleum Engineers Annual Technical Conference and Exhibition.[55] According to the report, drill cuttings are stored and transported through steel pipes and tanks—which the radiation cannot penetrate. However, improper transport and handling of drill cuttings could result in water contamination. For example, NORM

---

[53]GAO, *National Wildlife Refuges: Opportunities to Improve the Management and Oversight of Oil and Gas Activities on Federal Lands*, GAO-03-517 (Washington, D.C.: Aug. 28, 2003).

[54]Naturally occurring radioactive materials (NORM) are present at varying degrees in virtually all environmental media, including rocks and soils. According to a DOE report, human exposure to radiation comes from a variety of sources, including naturally occurring radiation from space, medical sources, consumer products, and industrial sources. Normal disturbances of NORM-bearing rock formations by activities such as drilling do not generally pose a threat to workers, the general public or the environment, according to studies and publications we reviewed.

[55]J. Daniel Arthur, Brian Bohm, David Cornue. "Environmental Considerations of Modern Shale Gas Development" (presented at the Society of Petroleum Engineers Annual Technical Conference and Exhibition, New Orleans, Louisiana, October 2009).

concentrations can build up in pipes and tanks, if not properly disposed, and the general public or water could come into contact with them, according to an EPA fact sheet.[56]

The chemical additives in fracturing fluid, if not properly handled, also poses a risk to water quality if they come into contact with surface water or groundwater. Some additives used in fracturing fluid are known to be toxic, but data are limited for other additives. For example, according to reports we reviewed, operators may include diesel fuel—a refinery product that consists of several components, possibly including some toxic impurities such as benzene and other aromatics—as a solvent and dispersant in fracturing fluid. While some additives are known to be toxic, less is known about potential adverse effects on human health in the event that a drinking water aquifer was contaminated as a result of a spill or release of fracturing fluid, according to the 2011 New York Department of Environmental Conservation's Supplemental Generic Environmental Impact Statement. This is largely because the overall risk of human health effects occurring from hydraulic fracturing fluid would depend on whether human exposure occurs, the specific chemical additives being used, and site-specific information about exposure pathways and environmental contaminant levels.

The produced water and fracturing fluids returned during the flowback process contain a wide range of contaminants and pose a risk to water quality, if not properly managed.[57] Most of the contaminants occur naturally, but some are added through the process of drilling and hydraulic fracturing. In January 2012, we reported that the range of contaminants found in produced water can include,[58] but is not limited to

- salts, which include chlorides, bromides, and sulfides of calcium, magnesium, and sodium;

---

[56]EPA, *Radioactive Waste from Oil and Gas Drilling*, EPA 402-F-06-038 (Washington, D.C.: April 2006).

[57]A 2009 report from DOE and the Groundwater Protection Council—a nonprofit organization whose members consist of state ground water regulatory agencies—estimates that from 30 percent to 70 percent of the original fluid injected returns to the surface.

[58]GAO-12-156.

- metals, which include barium, manganese, iron, and strontium, among others;

- oil, grease, and dissolved organics, which include benzene and toluene, among others;

- NORM; and

- production chemicals, which may include friction reducers to help with water flow, biocides to prevent growth of microorganisms, and additives to prevent corrosion, among others.

At high levels, exposure to some of the contaminants in produced water could adversely affect human health and the environment. For example, in January 2012, we reported that, according to EPA, a potential human health risk from exposure to high levels of barium is increased blood pressure.[59] From an environmental standpoint, research indicates that elevated levels of salts can inhibit crop growth by hindering a plant's ability to absorb water from the soil. Additionally, exposure to elevated levels of metals and production chemicals, such as biocides, can contribute to increased mortality among livestock and wildlife.

Operators must transport or store produced water prior to disposal. According to a 2012 University of Texas report, produced water temporarily stored in tanks (see fig. 12) or impoundments prior to treatment or disposal may be a source of leaks or spills, if not properly managed. The risk of a leak or spill is particularly a concern for surface impoundments as improper liners can tear, and impoundments can overflow.[60] For example, according to state regulators in North Dakota, in 2010 and 2011, impoundments overflowed during the spring melt season because operators did not move fluids from the impoundments—which

---

[59]GAO-12-156.

[60]The composition of pit lining depends on regulatory requirements, which vary from state to state.

were to be used for temporary storage—to a proper disposal site before the spring thaw.[61]

Figure 12: Storage Tank for Produced Water in the Barnett Shale

Source: GAO.

Unlike shale oil and gas formations, water permeates coalbed methane formations, and its pressure traps natural gas within the coal. To produce natural gas from coalbed methane formations, water must be extracted to lower the pressure in the formation so the natural gas can flow out of the coal and to the wellbore. In 2000, USGS reported that water extracted from coalbed methane formations is commonly saline and, if not treated

---

[61]In response, the state passed a new law that will significantly reduce the number of pits. Under the new law, operators can use pits for temporary storage of fluid from the flowback process but must drain and reclaim the pits no more than 72 hours after hydraulic fracturing is complete.

and disposed of properly, could adversely affect streams and threaten fish and aquatic resources.

According to several reports, handling and transporting toxic fluids or contaminants poses a risk of environmental contamination for all industries, not just oil and gas development; however, the large volume of fluids and contaminants—fracturing fluid, drill cuttings, and produced water—that is associated with the development of shale oil and gas poses an increased risk for a release to the environment and the potential for greater effects should a release occur in areas that might not otherwise be exposed to these chemicals.

Erosion

Oil and gas development, whether conventional or shale oil and gas, can contribute to erosion, which could carry sediments and pollutants into surface waters. Shale oil and gas development require operators to undertake a number of earth-disturbing activities, such as clearing, grading, and excavating land to create a pad to support the drilling equipment. If necessary, operators may also construct access roads to transport equipment and other materials to the site. As we reported in February 2005, as with other construction activities, if sufficient erosion controls to contain or divert sediment away from surface water are not established then surfaces are exposed to precipitation and runoff could carry sediment and other harmful pollutants into nearby rivers, lakes, and streams.[62] For example, in 2012, the Pennsylvania Department of Environmental Protection concluded that an operator in the Marcellus Shale did not provide sufficient erosion controls when heavy rainfall in the area caused significant erosion and contamination of a nearby stream from large amounts of sediment.[63] As we reported in February 2005, sediment clouds water, decreases photosynthetic activity, and destroys organisms and their habitat.

---

[62]GAO, *Storm Water Pollution: Information Needed on the Implications of Permitting Oil and Gas Construction Activities*, GAO-05-240 (Washington, D.C.: Feb. 9, 2005).

[63]In response, the state required the operator to install silt fences, silt socks, gravel surfacing of the access road, and a storm water capture ditch.

Underground Migration

According to a number of studies and publications we reviewed, underground migration of gases and chemicals poses a risk of contamination to water quality.[64] Underground migration can occur as a result of improper casing and cementing of the wellbore as well as the intersection of induced fractures with natural fractures, faults, or improperly plugged dry or abandoned wells. Moreover, there are concerns that induced fractures can grow over time and intersect with drinking water aquifers. Specifically:

Improper casing and cementing. A well that is not properly isolated through proper casing and cementing could allow gas or other fluids to contaminate aquifers as a result of inadequate depth of casing,[65] inadequate cement in the annular space around the surface casing, and ineffective cement that cracks or breaks down under the stress of high pressures. For example, according to a 2008 report by the Ohio Department of Natural Resources, a gas well in Bainbridge, Ohio, was not properly isolated because of faulty sealing, allowing natural gas to build up in the space around the production casing and migrate upward over about 30 days into the local aquifer and infiltrating drinking water wells.[66] The risk of contamination from improper casing and cementing is not unique to the development of shale formations. Casing and cementing practices also apply to conventional oil and gas development. However, wells that are hydraulically fractured have some unique aspects. For example, hydraulically fractured wells are commonly exposed to higher pressures than wells that are not hydraulically fractured. In addition, hydraulically fractured wells are exposed to high pressures over a longer period of time as fracturing is conducted in multiple stages, and wells may be refractured multiple times—primarily to extend the economic life of the well when production declines significantly or falls below the estimated reservoir potential.

---

[64]Methane can occur naturally in shallow bedrock and unconsolidated sediments and has been known to naturally seep to the surface and contaminate water supplies, including water wells. Methane is a colorless, odorless gas and is generally considered nontoxic, but there could be an explosive hazard if gas is present in significant volumes and the water well is not properly vented.

[65]The depth for casing and cementing may be determined by state regulations.

[66]Ohio Department of Natural Resources, *Report on the Investigation of the Natural Gas Invasion of Aquifers in Bainbridge Township of Geauga County, Ohio* (September 2008).

Natural fractures, faults, and abandoned wells. If shale oil and gas development activities result in connections being established with natural fractures, faults, or improperly plugged dry or abandoned wells, a pathway for gas or contaminants to migrate underground could be created—posing a risk to water quality. These connections could be established through either induced fractures intersecting directly with natural fractures, faults, or improperly plugged dry or abandoned wells or as a result of improper casing and cementing that allow gas or other contaminants to make such connections. In 2011, the New York State Department of Environmental Conservation reported that operators generally avoid development around known faults because natural faults could allow gas to escape, which reduces the optimal recovery of gas and the economic viability of a well. However, data on subsurface conditions in some areas are limited. Several studies we reviewed report that some states are unaware of the location or condition of many old wells. As a result, operators may not be fully aware of the location of abandoned wells and natural fractures or faults.

Fracture growth. A number of such studies and publications we reviewed report that the risk of induced fractures extending out of the target formation into an aquifer—allowing gas or other fluids to contaminate water—may depend, in part, on the depth separating the fractured formation and the aquifer. For example, according to a 2012 Bipartisan Policy Center report, [67] the fracturing process itself is unlikely to directly affect freshwater aquifers because fracturing typically takes place at a depth of 6,000 to 10,000 feet, while drinking water tables are typically less than 1,000 feet deep. [68] Fractures created during the hydraulic fracturing process are generally unable to span the distance between the targeted shale formation and freshwater bearing zones. According to a 2011 industry report, fracture growth is stopped by natural subsurface barriers

---

[67]Bipartisan Policy Center, *Shale Gas: New Opportunities, New Challenges* (Washington, D.C.: January 2012).

[68]Some coalbed methane formations are much closer to drinking water aquifers than are shale formations. In 2004, EPA reviewed incidents of drinking water well contamination believed to be associated with hydraulic fracturing in coalbed methane formations. EPA found no confirmed cases linked to the injection of fracturing fluid or subsequent underground movement of fracturing fluids. The report states that, although thousands of coalbed methane formations are fractured annually, EPA did not find confirmed evidence that drinking water wells had been contaminated by the hydraulic fracturing process.

GAO-12-732 Shale Oil and Gas Development

and the loss of hydraulic fracturing fluid.[69] When a fracture grows, it conforms to a general direction set by the stresses in the rock, following what is called fracture direction or orientation. The fractures are most commonly vertical and may extend laterally several hundred feet away from the well, usually growing upward until they intersect with a rock of different structure, texture, or strength. These are referred to as seals or barriers and stop the fracture's upward or downward growth. In addition, as the fracturing fluid contacts the formation or invades natural fractures, part of the fluid is lost to the formation. The loss of fluids will eventually stop fracture growth according to this industry report.

From 2001 through 2010, an industry consulting firm monitored the upper and lower limits of hydraulically induced fractures relative to the position of drinking water aquifers in the Barnett and Eagle Ford Shale, the Marcellus Shale, and the Woodford Shale.[70] In 2011, the firm reported that the results of the monitoring show that even the highest fracture point is several thousand feet below the depth of the deepest drinking water aquifer. For example, for over 200 fractures in the Woodford Shale, the typical distance between the drinking water aquifer and the top of the fracture was 7,500 feet, with the highest fracture recorded at 4,000 feet from the aquifer. In another example, for the 3,000 fractures performed in the Barnett Shale, the typical distance from the drinking water aquifer and the top of the fracture was 4,800 feet, and the fracture with the closest distance to the aquifer was still separated by 2,800 feet of rock. Table 4 shows the relationship between shale formations and the depth of treatable water in five shale gas plays currently being developed.

[69]George E. King, Apache Corporation, "Explaining and Estimating Fracture Risk: Improving Fracture Performance in Unconventional Gas and Oil Wells" (presented at the Society of Petroleum Engineers Hydraulic Fracturing Conference, The Woodlands, Texas, February 2012).

[70]Kevin Fisher, Norm Warpinski, Pinnacle—A Haliburton Service, "Hydraulic Fracture-Height Growth: Real Data" (presented at the Society of Petroleum Engineers Technical Conference and Exhibition, Denver, Colorado, October 2011).

Table 4: Shale Formation and Treatable Water Depth

Distance in feet

Shale play	Depth to shale	Depth to base of treatable water	Distance between shale and base of treatable water
Barnett	6,500- 8,500	1,200	5,300- 7,300
Fayetteville	1,000- 7,000	500	500- 6,500
Haynesville	10,500- 13,500	400	10,100- 13,100
Marcellus	4,000- 8,500	850	2,125- 7,650
Woodford	6,000- 11,000	400	5,600- 10,600

Source: GAO analysis of data presented in a report prepared at the request of the DOE.

Note: Depths to base of treatable water are approximate. According to the report, the depth to base of treatable water was based on data from state oil and gas agencies and state geological survey data.

Several government, academic, and nonprofit organizations evaluated water quality conditions or groundwater contamination incidents in areas experiencing shale oil and gas development. Among the studies and publications we reviewed that discuss the potential contamination of drinking water from the hydraulic fracturing process in shale formations are the following:

- In 2011, the Center for Rural Pennsylvania analyzed water samples taken from 48 private water wells located within about 2,500 feet of a shale gas well in the Marcellus Shale.[71] The analysis compared predrilling samples to postdrilling samples to identify any changes to water quality. The analysis showed that there were no statistically significant increases in pollutants prominent in drilling waste fluids—such as total dissolved solids, chloride, sodium, sulfate, barium, and strontium—and no statistically significant increases in methane. The study concluded that gas well drilling had not had a significant effect on the water quality of nearby drinking water wells.

- In 2011, researchers from Duke University studied shale gas drilling and hydraulic fracturing and the potential effects on shallow groundwater systems near the Marcellus Shale in Pennsylvania and the Utica Shale in New York. Sixty drinking water samples were collected in Pennsylvania and New York from bedrock aquifers that

---

[71]The Center for Rural Pennsylvania, *The Impact of Marcellus Gas Drilling on Rural Drinking Water Supplies* (Harrisburg, Pennsylvania: October 2011).

overlie the Marcellus or Utica Shale formations—some from areas with shale gas development and some from areas with no shale gas development.[72] The study found that methane concentrations were detected generally in 51 drinking water wells across the region—regardless of whether shale gas drilling occurred in the area—but that concentrations of methane were substantially higher closer to shale gas wells. However, the researchers reported that a source of the contamination could not be determined. Further, the researchers reported that they found no evidence of fracturing fluid in any of the samples.

- In 2011, the Ground Water Protection Council evaluated state agency groundwater investigation findings in Texas and categorized the determinations regarding causes of groundwater contamination resulting from the oil and gas industry.[73] During the study period—from 1993 through 2008—multistaged hydraulic fracturing stimulations were performed in over 16,000 horizontal shale gas wells. The evaluation of the state investigations found that there were no incidents of groundwater contamination caused by hydraulic fracturing.

In addition, regulatory officials we met with from eight states—Arkansas, Colorado, Louisiana, North Dakota, Ohio, Oklahoma, Pennsylvania, and Texas—told us that, based on state investigations, the hydraulic fracturing process has not been identified as a cause of groundwater contamination within their states.

A number of studies discuss the potential contamination of water from the hydraulic fracturing process in shale formations. However, according to several studies we reviewed, there are insufficient data for predevelopment (or baseline) conditions for groundwater. Without data to compare predrilling conditions to postdrilling conditions, it is difficult to determine if adverse effects were the result of oil and gas development, natural occurrences, or other activities. In addition, while researchers

---

[72]Stephen G. Osborn, Avner Vengosh, Nathaniel R. Warner, and Robert B. Jackson, "Methane Contamination of Drinking Water Accompanying Gas-well Drilling and Hydraulic Fracturing," *Proceedings of the National Academy of Science* 108, no. 20 (2011).

[73]Ground Water Protection Council, *State Oil and Gas Agency Groundwater Investigations And Their Role in Advancing Regulatory Reforms: A Two-State Review: Ohio and Texas* (Oklahoma City, Oklahoma: August 2011).

have evaluated fracture growth, the widespread development of shale oil and gas is relatively new. As such, little data exist on (1) fracture growth in shale formations following multistage hydraulic fracturing over an extended time period, (2) the frequency with which refracturing of horizontal wells may occur, (3) the effect of refracturing on fracture growth over time,[74] and (4) the likelihood of adverse effects on drinking water aquifers from a large number of hydraulically fractured wells in close proximity to each other.

Ongoing Studies Related to Water Quality

Ongoing studies by federal agencies, industry groups, and academic institutions are evaluating the effects of hydraulic fracturing on water resources so that, over time, better data and information about these effects should become available to policymakers and the public. For example, EPA's Office of Research and Development initiated a study in January 2010 to examine the potential effects of hydraulic fracturing on drinking water resources. According to agency officials, the agency anticipates issuing a progress report in 2012 and a final report in 2014. EPA is also conducting an investigation to determine the presence of groundwater contamination within a tight sandstone formation being developed for natural gas near Pavillion, Wyoming, and, to the extent possible, identify the source of the contamination. In December 2011, EPA released a draft report outlining findings from the investigation. The report is not finalized, but the agency indicated that it had identified certain constituents in groundwater above the production zone of the Pavillion natural gas wells that are consistent with some of the constituents used in natural gas well operations, including the process of hydraulic fracturing. DOE researchers are also testing the vertical growth of fractures during hydraulic fracturing to determine whether fluids can travel thousands of feet through geologic faults into water aquifers close to the surface.

Land and Wildlife

Oil and gas development, whether conventional or shale oil and gas, poses a risk to land resources and wildlife habitat as a result of constructing, operating, and maintaining the infrastructure necessary to develop oil and gas; using toxic chemicals; and injecting waste products underground.

---

[74]According to research presented in the New York Department of Environmental Conservation's Supplemental Generic Environmental Impact Statement, refracturing can restore the original fracture height and length, and can often extend the fracture length beyond the original fracture dimensions.

Habitat Degradation

According to studies and publications we reviewed, development of oil and gas, whether conventional or shale oil and gas, poses a risk to habitat from construction activities. Specifically, clearing land of vegetation and leveling the site to allow access to the resource, as well as construction of roads, pipelines, storage tanks, and other infrastructure needed to extract and transport the resource can fragment habitats.[75] In August 2003, we reported that oil and gas infrastructure on federal wildlife refuges can reduce the quality of habitat by fragmenting it.[76] Fragmentation increases disturbances from human activities, provides pathways for predators, and helps spread nonnative plant species.

In addition, spills of oil, gas, or other toxic chemicals have harmed wildlife and habitat. Oil and gas can injure or kill wildlife by destroying the insulating capacity of feathers and fur, depleting oxygen available in water, or exposing wildlife to toxic substances. Long-term effects of oil and gas contamination on wildlife are difficult to determine, but studies suggest that effects of exposure include reduced fertility, kidney and liver damage, immune suppression, and cancer. In August 2003, we reported that even small spills may contaminate soil and sediments if they occur frequently.[77] Further, noise and the presence of new infrastructure associated with shale gas development may also affect wildlife. A study by the Houston Advanced Research Center and the Nature Conservancy investigated the effects of noise associated with gas development on the Attwater's Prairie Chicken—an endangered species. The study explored how surface disruptions, particularly construction of a rig and noise from diesel generators would affect the animal's movement and habitat.[78] The results of the study found that the chickens were not adversely affected by the diesel engine generator's noise but that the presence of the rig caused the animals to temporarily disperse and avoid the area.

---

[75]Habitat fragmentation occurs when a network of roads and other infrastructure is constructed in previously undeveloped areas.

[76]GAO-03-517.

[77]GAO-03-517.

[78]James F. Bergan, Richard Haut, Jared Judy, and Liz Price. "Living In Harmony—Gas Production and the Attwater's Prairie Chicken" (presented at the Society of Professional Engineers Annual Technical Conference, Florence, Italy, September 2010).

A number of studies we reviewed identified risks to habitat and wildlife as a result of shale oil and gas activities. However, because shale oil and gas development is relatively new in some areas, the long-term effects—after operators are to have restored portions of the land to predevelopment conditions—have not been evaluated. Without these data, the cumulative effects of shale oil and gas development on habitat and wildlife are largely unknown.

Induced Seismicity

According to several studies and publications we reviewed, the hydraulic fracturing process releases energy deep beneath the surface to break rock but the energy released is not large enough to trigger a seismic event that could be felt on the surface. However, a process commonly used by operators to dispose of waste fluids—underground injection—has been associated with earthquakes in some locations. For example, a 2011 Oklahoma Geological Survey study reported that underground injection can induce seismicity. In March 2012, the Ohio Department of Natural Resources reported that "there is a compelling argument" that the injection of produced water into underground injection wells was the cause of the 2011 earthquakes near Youngstown, Ohio. In addition, the National Academy of Sciences released a study in June 2012 that concluded that underground injection of wastes poses some risk for induced seismicity, but that very few events have been documented over the past several decades relative to the large number of disposal wells in operation.

The available research does not identify a direct link between hydraulic fracturing and increased seismicity, but there could be an indirect effect to the extent that increased use of hydraulic fracturing produces increased amounts of water that is disposed of through underground injection. In addition, according to the National Academy of Science's 2012 report, accurately predicting magnitude or occurrence of seismic events is generally not possible, in part, because of a lack of comprehensive data on the complex natural rock systems at energy development sites.

Extent of Risks Is Unknown and Depends on Many Factors

The extent and severity of environmental and public health risks identified in the studies and publications we reviewed may vary significantly across shale basins and also within basins because of location- and process-specific factors, including the location and rate of development; geological characteristics, such as permeability, thickness, and porosity of the formations in the basin; climatic conditions; business practices; and regulatory and enforcement activities.

Location and rate of development. The location of oil and gas operations and the rate of development can affect the extent and severity of environmental and public health risks. For example, as we reported in October 2010, while much of the natural gas that is vented and flared is considered to be unavoidably lost, certain technologies and practices can be applied throughout the production process to capture some of this gas, according to the oil and gas industry and EPA. The technologies' technical and economic feasibility varies and sometimes depends on the location of operations. For example, some technologies require a substantial amount of electricity, which may be less feasible for remote production sites that are not on the electrical grid. In addition, the extent and severity of environmental risks may vary based on the location of oil and gas wells. For example, in areas with high population density that are already experiencing challenges adhering to federal air quality limits, increases in ozone levels because of emissions from oil and gas development may compound the problem.

Geological characteristics. Geological characteristics can affect the extent and severity of environmental and public health risks associated with shale oil and gas development. For example, geological differences between tight sandstone and shale formations are important because, unlike shale, tight sandstone has enough permeability to transmit groundwater to water wells in the region. In a sense, the tight sandstone formation acts as a reservoir for both natural gas and for groundwater. In contrast, shale formations are typically not permeable enough to transmit water and are not reservoirs for groundwater. According to EPA officials, hydraulic fracturing in a tight sandstone formation that is a reservoir for both natural gas and groundwater poses a greater risk of contamination than the same activity in a deep shale formation.

Climatic conditions. Climatic factors, such as annual rainfall and surface temperatures, can also affect the environmental risks for a specific region or area. For example, according to a 2007 study funded by DOE, average rainfall amounts can be directly related to soil erosion.[79] Specifically,

---

[79]ALL Consulting and the Interstate Oil and Gas Compact Commission, *Improving Access to Onshore Oil and Gas Resources on Federal Lands* (a special report prepared at the request of the U.S. Department of Energy National Energy and Technology Laboratory, March 2007).

areas with higher precipitation levels may be more susceptible to soil compaction and rutting during the well pad construction phase. In another example, risk of adverse effects from exposures to toxic air contaminants can vary substantially between drilling sites, in part, because of the specific mix of emissions and climatic conditions that affect the transport and dispersion of emissions. Specifically, wind speed and direction, temperature, as well as other climatic conditions, can influence exposure levels of toxic air contaminants. For example, according to a 2012 study from the Sustainable Investments Institute and the Investor Responsibility Research Center Institute, the combination of air emissions from gas operations, snow on the ground, bright sunshine, and temperature inversions during winter months have contributed to ozone creation in Sublette County, Wyoming.[80]

Business practices. A number of studies we reviewed indicate that some adverse effects from shale oil and gas development can be mitigated through the use of technologies and best practices. For example, according to standards and guidelines issued jointly by the Departments of the Interior and Agriculture, mitigation techniques, such as fencing and covers, should be used around impoundments to prevent livestock or wildlife from accessing fluids stored in the impoundments.[81] In another example, EPA's Natural Gas STAR program has identified over 80 technologies and practices that can cost effectively reduce methane emissions, a potent greenhouse gas, during oil and gas development. However, the use of these technologies and business practices are typically voluntary and rely on responsible operators to ensure that necessary actions are taken to prevent environmental contamination. Further, the extent to which operators use these mitigating practices is unknown and could be particularly challenging to identify given the significant increase in recent years in the development of shale oil and gas by a variety of operators, both large and small.

Regulatory and enforcement activities. Potential changes to the federal, state, and local regulatory environment will affect operators' future

---

[80]Susan Williams, "Discovering Shale Gas: An Investor Guide to Hydraulic Fracturing," Sustainable Investments Institute and Investor Responsibility Research Center Institute (New York, NY: February 2012).

[81]United States Department of the Interior and United States Department of Agriculture. *Surface Operating Standards and Guidelines for Oil and Gas Exploration and Development.* BLM/WO/ST-06/021+3071/REV 07 (Denver, CO: 2007).

activities and can therefore affect the risks or level of risks associated with shale oil and gas development. Shale oil and gas development is regulated by multiple levels of government—including federal, state, and local. Many of the laws and regulations applicable to shale oil and gas development were put in place before the increase in operations that has occurred in the last few years, and various levels of government are evaluating and, in some cases, revising laws and regulations to respond to the increase in shale oil and gas development. For example, in April 2012, EPA promulgated New Source Performance Standards for the oil and gas industry that, when fully phased-in by 2015, will require emissions reductions at new or modified oil and gas well sites, including wells using hydraulic fracturing. Specifically, these new standards, in part, focus on reducing the venting of natural gas and volatile organic compounds during the flowback process. In addition, areas without prior experience with oil and gas development are just now developing new regulations. These governments' effectiveness in implementing and enforcing this framework will affect future activities and the level of associated risk.

Agency Comments

We provided a draft of this report to the Department of Energy, the Department of the Interior, and the Environmental Protection Agency for review and comment. We received technical comments from Interior's Assistant Secretary, Policy, Management, and Budget, and from Environmental Protection Agency officials, which we have incorporated as appropriate. In an e-mail received August 27, 2012, the Department of Energy liaison stated the agency had no comments on the report.

As agreed with your offices, unless you publicly announce the contents of this report earlier, we plan no further distribution until 30 days from the report date. At that time, we will send copies of this report to the appropriate congressional committees, the Secretary of Energy, the Secretary of the Interior, the EPA Administrator, and other interested parties. In addition, the report will be available at no charge on the GAO website at http://www.gao.gov.

If you or your staff members have any questions about this report, please contact me at (202) 512-3841 or ruscof@gao.gov. Contact points for our Offices of Congressional Relations and Public Affairs may be found on the last page of this report. GAO staff who made key contributions to this report are listed in appendix IV.

Frank Rusco
Director, Natural Resources and Environment

List of Requesters

The Honorable Barbara Boxer
Chairman
Committee on Environment and Public Works
United States Senate

The Honorable Sheldon Whitehouse
Chairman
Subcommittee on Oversight
Committee on Environment and Public Works
United States Senate

The Honorable Benjamin L. Cardin
Chairman
Subcommittee on Water and Wildlife
Committee on Environment and Public Works
United States Senate

The Honorable Henry A. Waxman
Ranking Member
Committee on Energy and Commerce
House of Representatives

The Honorable Edward J. Markey
Ranking Member
Committee on Natural Resources
House of Representatives

The Honorable Diana DeGette
Ranking Member
Subcommittee on Oversight and Investigations
Committee on Energy and Commerce
House of Representatives

The Honorable Robert P. Casey, Jr.
United States Senate

Appendix I: Scope and Methodology

Our objectives for this review were to determine what is known about (1) the size of shale oil and gas resources in the United States and the amount produced from 2007 through 2011—the years for which data were available—and (2) the environmental and public health risks associated with development of shale oil and gas.

To determine what is known about the size of shale oil and gas resources, we collected data from federal agencies, state agencies, private industry, and academic organizations. Specifically, to determine what is known about the size of these resources, we obtained information for technically recoverable and proved reserves estimates for shale oil and gas from the Energy Information Administration (EIA), the U.S. Geological Survey (USGS), and the Potential Gas Committee—a nongovernmental organization composed of academic and industry officials. We interviewed key officials about the assumptions and methodologies used to estimate the resource size. Estimates of proved reserves of shale oil and gas are based on data provided to EIA by operators. In addition to the estimates provided by these three organizations, we also obtained and presented technically recoverable shale oil and gas estimates from two private organizations—IHS Inc., and ICF International—and one national advisory committee representing the views of the oil and gas industry and other stakeholders—the National Petroleum Council. For all estimates we report, we conducted a review of the methodologies used in these estimates for fatal flaws; we did not find any fatal flaws in these methodologies.

To determine what is known about the amount of produced shale oil and gas from 2007 through 2011, we obtained data from EIA—the federal agency responsible for estimating and reporting this and other energy information. EIA officials provided us with estimated oil and gas production data, including data estimating shale oil and gas estimates from states and two private firms—HPDI, LLC and Lippman Consulting, Inc. To assess the reliability of these data, we examined EIA's published methodology for collecting this information and interviewed key EIA officials regarding the agency's data collection and validation efforts. We also interviewed officials from three state agencies, representatives from five private companies, and researchers from three academic institutions who are familiar with these data and EIA's methodology and discussed the sources and reliability of the data. We determined that these data were sufficiently reliable for the purposes of this report.

To determine what is known about the environmental and public health risks associated with the development of shale oil and gas[1], we identified and reviewed more than 90 studies and other publications from federal agencies and laboratories, state agencies, local governments, the petroleum industry, academic institutions, environmental and public health groups, and other nongovernmental associations. The studies and publications we reviewed included scientific and industry periodicals, government-sponsored research, reports or other publications from nongovernmental organizations, and presentation materials. We identified these studies by conducting a literature search and by asking for recommendations during our interviews with stakeholders. For a number of studies, we interviewed the author or authors to discuss the study's findings and limitations, if any. We believe we have identified the key studies through our literature review and interviews, and that the studies included in our review have accurately identified potential risks for shale oil and gas development. However, given our methodology, it is possible that we may not have identified all of the studies with findings relevant to our objectives, and the risks we present may not be the only issues of concern. The widespread use of horizontal drilling and hydraulic fracturing to develop shale oil and gas is relatively new. Studying the effects of an activity and completing a formal peer-review process can take numerous months or years. Because of the relative short time frame for operations and the lengthy time frame for studying effects, we did not limit the review to peer-reviewed publications.

The risks identified in the studies and publications we reviewed cannot, at present, be quantified, and the magnitude of potential adverse affects or likelihood of occurrence cannot be determined for several reasons. First, it is difficult to predict how many or where shale oil and gas drilling operations may be constructed. Second, operators' use of effective best practices to mitigate risk may vary. Third, based on the studies we reviewed, there are relatively few that are based on evaluating predevelopment conditions to postdevelopment conditions—making it difficult to detect or attribute adverse changes to shale oil and gas development. In addition, changes to the federal, state, and local

---

[1]Operators may use hydraulic fracturing to develop oil and natural gas from formations other than shale. Specifically, coalbed and tight sand formations may rely on these practices, and some studies and publications we reviewed identified risks that can apply to these formations. However, many of the studies and publications we identified and reviewed focused primarily on the development of shale formations.

regulatory environment and the effectiveness in implementation and enforcement will affect operators' future activities. Moreover, risks of adverse events, such as spills or accidents, may vary according to business practices, which in turn, may vary across oil and gas companies making it difficult to distinguish between risks that are inherent to the development of shale oil and gas from risks that are specific to particular business practices.

To obtain additional perspectives on issues related to environmental and public health risks, we interviewed a nonprobability sample of stakeholders representing numerous agencies and organizations. (See app. II for a list of agencies and organizations contacted.) We selected these agencies and organizations to be broadly representative of differing perspectives regarding environmental and public health risks. In particular, we obtained views and information from federal officials from the Department of Energy's National Energy Technical Laboratory, the Department of the Interior's Bureau of Land Management and Bureau of Indian Affairs, and the Environmental Protection Agency; state regulatory officials from Arkansas, Colorado, Louisiana, North Dakota, Ohio, Oklahoma, Pennsylvania, and Texas; tribal officials from the Osage Nation; shale oil and gas operators; representatives from environmental and public health organizations; and other knowledgeable parties with experience related to shale oil and gas development, such as researchers from the Colorado School of Mines, the University of Texas, Oklahoma University, and Stanford University. The findings from our interviews with stakeholders and officials cannot be generalized to those we did not speak with.

We conducted this performance audit from November 2011 to September 2012 in accordance with generally accepted government auditing standards. Those standards require that we plan and perform the audit to obtain sufficient, appropriate evidence to provide a reasonable basis for our findings and conclusions based on our audit objectives. We believe that the evidence obtained provides a reasonable basis for our findings and conclusions based on our audit objectives.

Appendix II: List of Agencies and Organizations Contacted

Federal Agencies	Congressional Research Service Department of Energy's National Energy Technology Laboratory Department of Health and Human Services Department of the Interior's Bureau of Indian Affairs Department of the Interior's Bureau of Land Management Department of the Interior's U.S. Geological Survey Environmental Protection Agency
State Agencies	Arkansas Department of Environmental Quality Arkansas Oil and Gas Commission Colorado Oil and Gas Conservation Commission Louisiana Department of Natural Resources North Dakota Industrial Commission Ohio Department of Natural Resources Ohio Environmental Protection Agency Oklahoma Geological Survey Oklahoma Corporation Commission Texas Railroad Commission
Academic Institutions	Colorado School of Mines Oklahoma University Stanford University University of Texas at Arlington University of Texas Energy Center and Bureau of Economic Geology
Environmental Organizations	Clean Water Action Pennsylvania Earthworks Oil and Gas Accountability Project Environmental Defense Fund Subra Consulting Western Resource Advocates
Public Health Organizations	The Endocrine Disruption Exchange National Association of County and City Health Officials Southwest Pennsylvania Environmental Health Project
Industry	ALL Consulting American Exploration and Production Council American Petroleum Institute Apache Corporation

Chesapeake Energy
Colorado Oil and Gas Association
Devon Energy
Powell Shale Digest

Others

Ground Water Protection Council
Martin Consulting
Red River Watershed Management Institute
Osage Tribal Nation

Appendix III: Additional Information on USGS Estimates

The USGS estimates potential oil and gas resources in about 60 geological areas (called "provinces") in the United States. Since 1995, USGS has conducted oil and gas estimates at least once in all of these provinces; about half of these estimates have been updated since the year 2000 (see table 5). USGS estimates for an area are updated once every 5 years or more, depending on factors such as the importance of an area.

Table 5: USGS Estimates

Name of USGS province	Most recent assessment year
Northern Alaska	2006
Central Alaska	2004
Southern Alaska	2011
Western Oregon-Wash.	2009
Eastern Oregon-Wash.	2006
Northern Coastal	1995
Sonoma-Livermore	1995
Sacramento Basin	2006
San Joaquin Basin	2004
Central Coastal	1995
Santa Maria Basin	1995
Ventura Basin	1995
Los Angeles Basin	1995
Idaho-Snake River Downwarp	1995
Western Great Basin	1995
Eastern Great Basin	2004
Uinta-Piceance Basin	2002
Paradox Basin	1995
San Juan Basin	2002
Albuquerque-Sante Fe Rift	1995
Northern Arizona	1995
S. Ariz.-S.W. New Mexico	1995
South-Central New Mexico	1995
Montana Thrust Belt	2002
Central Montana	2001
Southwest Montana	1995
Hanna, Laramie, Shirley	2005

Name of USGS province	Most recent assessment year
Williston Basin (includes Bakken Shale Formation)	2008
Powder River Basin	2006
Big Horn Basin	2008
Wind River Basin	2005
Wyoming Thrust Belt	2004
Southwestern Wyoming	2002
Park Basins	1995
Denver Basin	2003
Las Animas Arch	1995
Raton Basin-Sierra Grande Uplift	2005
Palo Duro Basin	1995
Permian Basin (includes Barnett Shale)	2007
Bend Arch-Ft. Worth Basin	2004
Marathon Thrust Belt	1995
Western Gulf Coast (includes Eagle Ford Shale)	2011
East Texas Basin Province	2011
Louisiana-Mississippi Salt Basins Province	2011
Florida Peninsula	2000
Superior	1995
Cambridge Arch-Central Kansas	1995
Nemaha Uplift	1995
Forest City Basin	1995
Anadarko Basin	2011
Sedgwick Basin/Salina Basin	1995
Cherokee Platform	1995
Southern Oklahoma	1995
Arkoma Basin	2010
Michigan Basin	2005
Illinois Basin	2007
Black Warrior Basin	2002
Cincinnati Arch	1995
Appalachian Basin (includes Marcellus Shale)	2011
Blue Ridge Thrust Belt	1995
Piedmont	1995

Source: USGS.

Appendix IV: GAO Contact and Staff Acknowledgments

GAO Contact	Frank Rusco, (202) 512-3841 or ruscof@gao.gov
Staff Acknowledgments	In addition to the contact named above, Christine Kehr, Assistant Director; Lee Carroll; Nirmal Chaudhary; Cindy Gilbert; Alison O'Neill; Marietta Revesz, Dan C. Royer; Jay Spaan; Kiki Theodoropoulos; and Barbara Timmerman made key contributions to this report.

www.ingramcontent.com/pod-product-compliance
Lightning Source LLC
Chambersburg PA
CBHW081219170526
45165CB00009B/2879